W9-DDG-145

BASIC PRINCIPLES OF LEARNING

BASIC PRINCIPLES OF LEARNING

Roger M. Tarpy
Bucknell University

Lyle E. Bourne, Jr., Consulting Editor
University of Colorado at Boulder

Scott, Foresman and Company

Glenview, Illinois

Dallas, Tex.   Oakland, N.J.   Palo Alto, Cal.   Tucker, Ga.   Brighton, England

ACKNOWLEDGEMENTS FOR FIGURES

(2-3) Lubow, R. E. "Latent inhibition: Effects of frequency of nonreinforced pre-exposure of the CS," *Journal of Comparative Physiological Psychology*, 1965, 60, 454-457. Figure 2 on p. 456. Reprinted by permission of the American Psychological Association and the author. (2-4) Kimmel, H. D. "Instrumental inhibitory factors in classical conditioning." In: W. F. Prokasy (ed.) *Classical Conditioning*. New York: Appleton-Century-Crofts, Inc., 1965. Copyright © 1965 by Appleton-Century-Crofts. Reprinted by permission of Prentice-Hall, Inc., and the author. (2-5) Fitzgerald, R. D., and Taylor, T. J. "Trace and delayed heart-rate conditioning in rats as a function of US intensity," *Journal of Comparative Physiological Psychology*, 1970, 70, 242-253. Reprinted by permission of the American Psychological Association and the author. (2-6) Coppock, W. J. "Pre-extinction in sensory preconditioning," *Journal of Experimental Psychology*, 1958, 55, 213-219. Figure 2 on p. 216. Reprinted by permission of the American Psychological Association and the author. (2-8) Kalat, J. W., and Rozin, P. "Role of interference in taste-aversion learning," *Journal of Comparative Physiological Psychology*, 1971, 77, 53-58. Figure 1 on p. 54. Reprinted by permission of the American Psychological Association and the author. (2-9) Wagner, A. R. "Stimulus selection and a 'Modified Continuity Theory'" in: *The Psychology of Learning and Motivation*, Volume 3, edited by G. H. Bower and J. T. Spence. Copyright © 1969. Reprinted by permission of Academic Press, Inc. (3-3) Thomas, J. R. "Fixed-ratio punishment by time-out of concurrent variable-interval behavior," *Journal of the Experimental Analysis of Behavior*, 1968, 11, 609-616. Figure 1 on p. 611. Copyright © 1968 by the Society for the Experimental Analysis of Behavior, Inc. Reprinted by permission. (3-4) Kintsch, W. "Runway performance as a function of drive strength and magnitude of reinforcement," *Journal of Comparative Physiological Psychology*, 1962, 55, 882-887. Figure 2 on p. 883. Reprinted by permission of the American Psychological Association and the author. (3-5) Bower, G H. "A contrast effect in differential conditioning," *Journal of Experimental Psychology*, 1961, 62, 196-199. Figure 1 on p. 197. Reprinted by permission of the American Psychological Association and the author. (3-7) McHose, J. H., and Tauber, L. "Changes in delay of reinforcement in simple instrumental conditioning," *Psychonomic Science*, 1972, 27, 291-292. Figure 1 on p. 291. Reprinted by permission of the Psychonomic Society, Inc., and the author. (3-8) Zaretsky, H. H. "Runway performance during extinction as a function of drive and incentive," *Journal of Comparative Physiological Psychology*, 1965, 60, 463-464. Figure 1 on p. 464. Reprinted by permission of the American Psychological Association and the author. (3-9) Trapold, M. A., and Fowler, H. "Instrumental escape performance as a function of the intensity of noxious stimulation," *Journal of Experimental Psychology*, 1960, 60, 323-326. Figure 2 on p. 325. Reprinted by permission of the American Psychological Association and the author. (3-10) Fowler, H., and Trapold, M. A. "Escape performance as a function of delay of reinforcement," *Journal of Experimental Psychology*, 1962, 63, 464-467. Figure 1 on p. 465. Reprinted by permission of the American Psychological Association and the author. (3-11) Tarpy, R. M., and Koster, E. D. "Stimulus facilitation of delayed-reward learning in the rat," *Journal of Comparative Physiological Psychology*, 1970, 71, 147-151. Figure 1 on p. 148. Reprinted by permission of the American Psychological Association and the author. (3-12) Felton, M., and Lyon, D. O. "The post-reinforcement pause," *Journal of the Experimental Analysis of Behavior*, 1966, 9, 131-134. Figure 3 on p. 133. Copyright © 1966 by the Society for the Experimental Analysis of Behavior, Inc. Reprinted by permission. (3-13) From *Schedules of Reinforcement* by Ferster, C. B., and Skinner, B. F. Copyright © 1957 by Appleton-Century-Crofts, Inc. Reprinted by permission of Prentice-Hall, Inc., and C. B. Ferster. (3-14) Miller, N. E., and Banuazizi, A. "Instrumental learning by curarized rats of a specific visceral response, intestinal, or cardiac," *Journal of Comparative Physiological Psychology*, 1968, 65, 1-7. Figures 3 and 4 on p. 5. Reprinted by permission of the American Psychological Association and the author. (4-2) Annau, Z., and Kamin, L. J. "The conditioned response as a function of intensity of the US," *Journal of Comparative Physiological Psychology*, 1961, 54, 428-432. Figure 1 on p. 430. Reprinted by permission of the American Psychological Association and the author. (4-4) Martin, L. K., and Riess, D. "Effects of US intensity during previous discrete delay conditioning on conditioned acceleration during avoidance extinction," *Journal of Comparative Physiological Psychology*, 1969, 69, 196-200. Figure 1 on p. 198.

of the American Psychological Association and the author. (8-2) Kelleher, R. T., and Fry, W. T. "Stimulus functions in chained fixed-interval schedules," *Journal of the Experimental Analysis of Behavior*, 1962, 5, 167-173. Figure 5 on p. 169. Copyright © 1962 by the Society for the Experimental Analysis of Behavior, Inc. Reprinted by permission. (8-3) Miles, R. C. "The relative effectiveness of secondary reinforcers throughout deprivation and habit-strength parameters," *Journal of Comparative Physiological Psychology*, 1956, 49, 126-130. Figure 1 on p. 127. Reprinted by permission of the American Psychological Association and the author. (8-4) Saltzman, I. J. "Maze learning in the absence of primary reinforcement: A study of secondary reinforcement," *Journal of Comparative Physiological Psychology*, 1949, 42, 161-181. Figure 4 on p. 170. Reprinted by permission of the American Psychological Association and the author. (8-5) Jenkins, W. O. "A temporal gradient of derived reinforcement," *American Journal of Psychology*, 1950, 63, 237-243. Figure 1 on p. 240. Reprinted by permission of The University of Illinois Press and the author. (8-6) Tombaugh, T. N. "Secondary reinforcement and the partial reinforcement effect in the rat," *Journal of Comparative Physiological Psychology*, 1970, 71, 160-164. Figure 2 on p. 163. Reprinted by permission of the American Psychological Association and the author. (8-7) Dinsmoor, J. A. "A quantitative comparison of the discriminative and reinforcing functions of a stimulus," *Journal of Experimental Psychology*, 1950, 40, 458-472. Figure 1 on p. 464. Reprinted by permission of the American Psychological Association and the author. (8-8) Stein, L. "Secondary reinforcement established with subcortical stimulation," *Science*, 1958, 127, 466-467. Reprinted by permission of the American Association for the Advancement of Science and the author. (8-10) Marx, M. H., and Murphy, W. W. "Resistance to extinction as a function of the presentation of a motivating cue in the start box," *Journal of Comparative Physiological Psychology*, 1961, 54, 207-210. Figure 1 on p. 209. Reprinted by permission of the American Psychological Association and the author. (9-1), (9-2) Guttman, N., and Kalish, H. I. "Discriminability and stimulus generalization." *Journal of Experimental Psychology*, 1956, 51, 79-88. Copyright © 1956 by the American Psychological Association. Reprinted by permission. (9-3) Friedman, H., and Guttman, N. "Further analysis of the various effects of discrimination training on stimulus generalization gradients," In: *Stimulus Generalization*, D. I. Mostofsky, ed., Stanford, Calif.: Stanford Univ. Press, 1965. Figure 8 on p. 265. Reprinted by permission of Stanford University Press. (9-5) Grice, G. R. "Investigations of response-mediated generalization," In: *Stimulus Generalization*, D. I. Mostofsky, ed., Stanford, Calif.: Stanford Univ. Press, 1965. Figure 1 on p. 374. Reprinted by permission of Stanford University Press. (9-6) Perkins, C. C., and Weyant, R. G. "The interval between training and test trials as a determiner of the slope of generalization gradients," *Journal of Comparative Physiological Psychology*, 1958, 51, 596-600. Copyright © 1958 by the American Psychological Association. Reprinted by permission. (9-7) Hanson, H. M. "Stimulus generalization following three-stimulus discrimination training," *Journal of Comparative Physiological Psychology*, 1961, 54, 181-185. Copyright © 1961 by the American Psychological Association. Reprinted by permission. (9-8) Hanson, H. M. "Effects of discrimination training on stimulus generalization," *Journal of Experimental Psychology*, 1959, 58, 321-334. Copyright © 1959 by the American Psychological Association. Reprinted by permission. (9-9) Kalish, H. I., and Haber, A. "Prediction of discrimination from generalization following variations in deprivation level," *Journal of Comparative Physiological Psychology*, 1965, 60, 125-128. Copyright © 1965 by the American Psychological Association. Reprinted by permission. (9-10) Harlow, H. F. "The formation of learning sets," *Psychological Review*, 1949, 56, 51-65. Copyright 1949 by the American Psychological Association. Reprinted by permission. (9-11) Warren, J. M. "Primate learning in comparative perspective." In: A. M. Schrier, H. F. Harlow, and F. Stollnitz (eds.), *Behavior of Nonhuman Primates: Modern Research Trends*. Copyright © 1965 by Academic Press and reprinted with their permission. (10-1) Coppock, H. W., and Chambers, R. M. "Reinforcement of position preference by automatic intravenous injections of glucose," *Journal of Comparative Physiological Psychology*, 1954, 47, 355-357. Copyright 1954 by the American Psychological Association. (10-2) "Predictions of the comparative reinforcement values of running and drinking," by D. Premack, *Science*, Vol. 139, pp. 1062-1063, March 1963. Copyright © 1963 by American Association for the Advancement of Science. Reprinted by permission of the American Association for the Advancement of Science and the author.

Preface

During the 1940s and 1950s, research in animal learning was one of the most popular and highly esteemed areas in psychology. The grand theories of behavior devised during those years, best exemplified in the work of Clark L. Hull, possessed vitality, precision, and breadth. In addition, the theories developed a viable technology and language within which learning principles were discovered and evaluated.

Such grand schemes, however, have recently faded from the field of animal learning, largely because they were ultimately unsuccessful in their account of the many vagaries of learning which have since proven to be exceedingly extensive and complex. The demise of such grand schemes was taken by many to be the death knell for learning research in general.

However, basic learning is far from being a vacuous or dead area. In fact, it has continued to persist as a dominant and important area of psychology. While other approaches have recently begun to flourish (physiological, ethological, cognitive), they have enriched and broadened the field of animal learning rather than replaced it. Learning theorists themselves have developed new perspectives and techniques, and the field is now healthier than ever.

This volume attempts to deal with these recent developments in basic animal learning. Approximately 65 percent of the cited references have been published since Kimble's (1961) superb book on learning, and of these, over 40 percent have been published within the last six years. Yet, this book also shows how and why these recent developments evolved from earlier formulations. Such an historical treatment is essential for it illustrates that learning research is a dynamic process, continually changing its theories on the basis of new evidence. What we currently believe to be true about learning reflects a collective judgment, or a consensus, formulated within the limits of available data rather than an end point in our inquiry. As is true with all texts, the judgments in this volume represent the interests and bias of the author, and no claim is made that different interpretations of the same material are not tenable.

I am deeply indebted to many students and colleagues who, directly or indirectly, have facilitated the writing of this book. Professor Lyle E. Bourne, Jr., provided constant support and guidance without which the book would have suffered enormously. Dr. Linda R. Warren and Mr. Fred L. Sawabini read the original manuscript and offered many perceptive criticisms for its improvement. Dr. George R. Jacobson exercised his impressive skill in writing during the final revisions, and whatever clarity exists is due to his fine efforts. Mrs. Dante Giusti, who typed the manuscript, was indispensable; her sanguine reactions to my scratchy handwriting were a constant source of amazement. I greatly appreciate the courtesy extended by the many publishers and authors who have allowed me to make generous use of their material. Finally, I would like to thank my family for their continuous encouragement throughout the many months spent in preparation of this volume.

Williamstown, Mass.

Roger M. Tarpy

Contents

Introduction

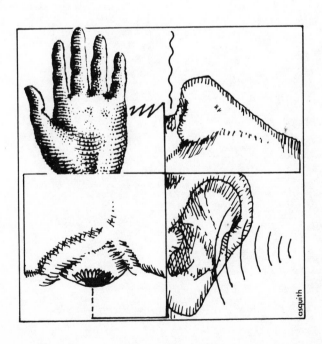

Psychologists study behavior, attempting to describe and understand it. Since much of behavior is learned, or acquired, the study of learning has come to occupy a central position in psychology. Furthermore, interest in learning is clearly not just academic. Most people have been taught to value the development of personal abilities and skills, and as a society, we have collectively emphasized education.

But how does one describe and understand learning? What does it mean to say a person has learned to speak Spanish, to use a typewriter, or to throw a baseball? How does a dog learning to jump through a hoop differ from a child "learning" to eat spinach? What do these various behaviors have in common? How is learning achieved, and how can the conditions under which we learn be improved?

Psychologists have been studying the learning process, both in and out of the laboratory, for about sixty years. In doing so, they have attempted to describe the kinds of responses that organisms are capable of learning, and to formulate the basic principles governing learning processes.

A Definition of Learning

Even though one can usually recognize instances of learning, it is difficult to adequately define it in a formal sense. Learning clearly represents a process by which an organism's behavior is changed. But not all change results from learning. It is necessary to distinguish more carefully between changes that are a consequence of learning and those changes due to other factors. For example, fatigue may alter behavior, but those changes are unrelated to learning. In a similar manner, changes in behavior due to maturation should not be classified as acquired, or learned. A child grows in stature until he can finally reach the handle and open a door. This change in behavior involves, to a large degree, physical maturation (growing tall) although learning may also play a role. It should be obvious that if such a behavior does involve learning, maturational factors, and, say, fatigue too, analysis of behavior becomes exceedingly complex.

Learning may be defined as a relatively permanent change in behavior which occurs as a result of experience. This definition, similar to one given by Kimble (1961), eliminates fatigue and motivational factors as possible causes for the change by using the phrase "relatively permanent." Maturational factors are excluded as causes too by stating that the change is due to "experience." Although this definition seems adequate, it does not actually specify *what* is learned (this issue is discussed in Chapter 10). For the present discussion, it is perhaps more useful simply to note instances of learning while leaving consideration of what is actually learned to a later section.

Historical Background

To understand learning research, it is helpful to consider briefly the origins of experimental psychology (for a complete discussion of the history of psychology, see Boring, 1957).

Throughout intellectual history, philosophers have been preoccupied with the mind-body distinction. Descartes felt that man's physical body was mechanistic, animallike, and involuntary, whereas his mind was spiritual, voluntary, and free. Man was comprised of mind and body; man's body existed on one level while his soul and mind existed on a separate level.

These concepts gave rise to a growing concern for understanding the internal dimensions of human existence. What was the mind; how does one measure and understand mental sensations and ideas which exist without physical form?

One answer was found in the brilliant work of Fechner, who attempted to measure mental sensations in terms of their physical antecedents. For example, loudness, which is an internal psychological sensation (mind), could be measured or scaled in terms of the quantity of sound energy (physical) which produced the sensation. This was done by stating how many decibels of sound intensity a person required before he could hear the sound, i.e., before a mental or psychological event took place. Similarly, a psychological sensation corresponding to a color stimulus could be measured according to physical characteristics of the stimulus, e.g., wavelength. The psychological event of, say, redness bears some relationship to the physical wavelength of the red stimulus. Thus, Fechner developed a way of relating mental events to physical events, of crudely measuring the psychological world.

Armed with this new technique, psychology became the science of consciousness or the mind, of mental events such as sensations and images. It was the task of such early psychologists as Titchener to study the structure of these events, the contents of man's experience. Consciousness had elements which, in turn, were described according to various attributes. For example, one basic unit or element of the mind was sensation. Each sensation, like redness, could be described according to about five attributes or characteristics: Its intensity, vividness, duration, and clarity were all dimensions or features of the sensation or mental activity. Thus, psychologists tried to create a taxonomy of mental experience, to account for the basic elements, and to describe what these elements were like.

The history of these attempts, of course, is varied and extensive. Several "schools" existed which differed in their theoretical emphasis. However, the two things which strongly characterized all of these psychologies were their common interest in studying the mind (consciousness, experience) and their experimental technique. The technique, introspection, involved an attempt to perceive in one's own experience the nature of that experience. It was a subjective evaluation of mental events by following systematic and objective guidelines. Introspectionists were trained to describe the "core" of their mental experience (the actual mental event) rather than the external stimulus. In other words, they were interested in making a description of the redness as a psychological event, rather than of the red stimulus itself. In summary then, consciousness (or mental elements) was the principal subject matter of psychologists, and they attempted to study this subjectively using the method of introspection.

Behaviorism

The most fundamental and far-reaching event took place in the early part of the twentieth century and dramatically changed the nature of psychological inquiry. The event was the behaviorist or Watsonian revolution. John B. Watson, a psychologist at Johns Hopkins University, spurned the techniques of introspection as well as the subject matter of previous psychologists. His claim was a simple and straightforward one: Consciousness was an internal and unobservable state which was not accessible to objective scientific inquiry. He proposed that the subject matter of psychology should be overt behavior which was observable and measurable. As a consequence, psychology could be objective and deterministic. Watson did not claim that consciousness did not exist. Instead, he simply stated that it was not a public event, accessible to objective study. What psychology had to study was overt behavior.

Psychology of Learning Today

Psychology today owes a great debt to Watson and his insight. Although strict behaviorism as proposed by Watson is not advocated by all psychologists, the vast majority of them have adopted Watson's basic philosophy of studying overt behavior systematically and objectively. Psychology today is behavioristic, certainly in the general sense of studying behavior systematically.

In general, the psychology of learning has evolved along with the behaviorist point of view attempting to discover the basic underlying principles of learned behavior. As in the natural sciences, learning psychologists have often studied limited systems or responses, or simple organisms, in an effort to discover learning principles that are common to all species or to all forms of learning. Thus a topic so complex as learned behavior may be reduced and analyzed in a simpler form.

Reducing the complexity of the system to a more basic, manageable level offers both advantages as well as disadvantages. For example, in studying the behavior of white rats, the experimenter can control the genetic makeup and previous history of the animal which, if not controlled, might seriously affect the result of the experiment and thereby alter the conclusions about learning. Clearly, such control is not always possible in humans.

Furthermore, most learning psychologists have believed that simple, arbitrary responses performed by rats are useful to study precisely because they are simple and arbitrary. The psychologist can manipulate various conditions in studying these basic behaviors and hopefully understand the underlying principles. Although this simple behavior differs in extent and kind from complex human behavior, it may offer a way of conceptualizing the principles of all learned behavior. Psychology is not really interested in the fact that dogs learn to jump through hoops or that rats learn to press levers; such behaviors are trivial. However, psychology is interested in the general characteristics of the behavior and in the variables (training procedures) which influence those characteristics. The relationships between these variables and behavior may be considered as the underlying principles of learning.

Once psychologists have begun to specify the basic units or principles of learned behavior, they may develop models and theories. Models are a set of principles, an account of the behavior; they represent or schematize the units and relationships in behavior in much the same way that a blueprint schematically represents the dimensions and elements of a building. Other sciences have models. For example, chemistry has the Law of Perfect Gases which notates a collection of principles concerning the behavior of gases in an idealized state although, in reality, no such gas exists. In essence then, a model unifies principles and relates the elements and functions of behavior.

There are some disadvantages and risks, however, in simplifying the system. The models of behavior which can be developed, in fact the principles of learning themselves, may be severely restricted to the organisms used in the research. Human learning may be not only more complex than rat learning, it may comply to different principles altogether. Of course there is no way to counter this criticism except to say that both notions are probably true: Human learning may involve unique principles which must be discovered by examining human behavior, yet there may also exist a variety of universal principles which are truly basic to learning and which apply to all species capable of learning.

Specific Historical Antecedents to Learning Research

Associationists

The psychology of learning evolved gradually and has its roots in several important historical movements. One of the earliest historical antecedents was the work of the British Empiricists, or Associationists. The prevailing philosophy of the time was that of Descartes who believed that ideas were innate: Man was born with ideas of motion, space, and other concepts, and experience simply filled in the details. The Empiricists, on the other hand, claimed that knowledge or ideas were derived from experience directly. Sensations gave rise to ideas; man's mental life developed directly from experience. This theory is most clearly documented in Locke's notion of *tabula rasa* (blank tablet). According to Locke, man is born without a preestablished mental life. His mind is basic "raw material" acted upon by experience.

The physical world of the senses totally determines the nature of man's mental life.

This point is important because the Empiricists, especially Hartley, developed laws determining the creation of knowledge or the development of mental ideas from experience. These Laws of Association accounted for the relationship between images and ideas; they were the "glue" in establishing a correspondence between the sensory, physical world and mental ideas.

One of these laws was the Law of Contiguity, which stated that sensations which occurred close in time to one another would give rise to an association of those events in the mind. If sensation A were closely followed by sensation B, a mental association between the two developed because they were contiguous, such that on later occasions the presentation of sensation A alone would evoke the memory of B. The Law of Contiguity thus postulated a mechanism by which mental (psychological) events were created, or in other terms, how associations were learned.

The Law of Contiguity was, in essence, the first theory of learning. In the late nineteenth century, Hermann Ebbinghaus studied these laws explicitly by examining the strength of individual associations. Using himself as his principal subject, he took nonsense syllables (e.g., BJK, QRV) which he thought were pure stimuli devoid of prior associations, and learned to connect one syllable with another. During a later memory test, he presented the first syllable and tried to recall the one associated with it. He experimented with different degrees of recitation and a variety of other conditions to see their effect on the memory of each association. Ebbinghaus was able to identify some important variables which contribute to memory, and many of his original findings are still valid today. More importantly, the Law of Contiguity was shown to be a major principle of association learning.

Darwin's Evolutionary Theory

A second major antecedent to the study of learning was Darwin's theory of evolution, in which he placed great emphasis on adaptive behavior whose function was to promote the continuation of the species. From this point of view, the classical problem of consciousness was irrelevant; rather what was important was an analysis of adaptive behavior. Learning results in adaptive behavior, and the ability to learn plays a major role in the preservation of many species to the extent that the animals learn about their environment—what dangers to avoid, where to seek food. The study of the way such adaptive behaviors develop thus is extraordinarily important because these behaviors account for a major factor in evolutionary theory.

A second important contribution of Darwin's theory to the study of learning was the notion of evolution itself. If all organisms evolved from simpler forms of life, a continuum or hierarchy of animal life exists. Despite differences in complexity, this notion of a continuum suggests that inferences may be made about, say, man from a study of monkeys or rats. Because of having the same underlying evolutionary basis, there is the possibility that learning principles will hold for species of different complexity. On a simpler level, the theory of evolution implies that man is not unique, and therefore his behavior may be explained by the same principles which account for the behavior of lower species.

Social Climate in America and Russia

Finally, an important historical condition giving rise to an interest in learning was the social climate in the United States and Russia. In these countries, but virtually nowhere else, learning research flourished. At the turn of the century, America was a burgeoning technological society. Thousands of immigrants were beginning to pour into the country, with the belief that great opportunities were available for those who could learn and develop skills. There was a keen sense of individualism, of hope in a new culture where status and wealth could be obtained according to one's abilities rather than according to one's inheritance. These notions are clearly documented by the emphasis placed on education by American society. In Europe, formal schooling, certainly at the university level, was too often available only to a privileged few. In contrast, in the United States, education was highly valued.

The situation was similar in Russia. The oppressive rule of the czars was overthrown in a revolution, and class structure was less rigidly defined by family status. Abilities which were acquired became the means to one's future.

Measurement of Learning

Learning Versus Performance

Earlier it was suggested that instances where learning takes place could be identified (as distinct from behavioral changes due to, say, fatigue) but that the "real" nature of learning, what was learned, would be left to a later chapter. Specifying when learning occurs but not identifying learning itself results in a distinction between learning and performance.

When a psychologist views any learning situation, whether it be a human learning to drive a car or a monkey learning to operate a slot machine, he is looking at overt behavior; the only objective measure he has to consider is performance. Learning, in contrast, refers to the underlying process which is believed to determine the performance, at least in part. The psychologist cannot directly observe learning but rather can only infer its presence and nature. Performance, however, is the expression of learning in behavioral terms, although performance may also be due to other variables such as maturation, or fatigue.

Some theorists have stated that since psychologists can never understand or observe learning directly, the distinction between learning and performance is a useless one. The two terms are operationally synonymous and indistinguishable. Since it would be impossible to discover principles of learning without measuring performance, these psychologists ignore the distinction and simply attempt to identify all the variables affecting performance. Under this scheme, no definition of learning could be given; rather learning would be regarded simply as performance, but not due to fatigue or other related variables. Other psychologists, however, feel that the distinction between learning and performance is not only useful but essential. While it is true that learning may be assessed only by measuring performance, there is growing suspicion that learning may involve different principles than performance.

A crucial demonstration of the learning-performance distinction would be to show that performance does not always reflect what was learned. Such a result has been shown in studies of latent learning (Blodgett, 1929; Tolman & Honzik, 1930; Thistlethwaite, 1951). For example, Blodgett had two groups of hungry rats; each was tested in a maze. The control group received food at the end of the maze and, as indicated by their errors in traversing the maze, showed gradual improvement, or learning, with continued training. The experimental group was not fed at the end of the maze, and consequently showed only a little improvement in their error scores. On the fourth trial, however, Blodgett began feeding the experimental group for the first time. Their error scores immediately decreased. In fact, the improvement in their performance occurred so quickly, so unlike the control group's performance, that it is difficult to believe that the change following trial four was due to learning. Rather, it appears that the experimental group was learning the maze all along but just wasn't performing in a manner that demonstrated their learning. However, when fed, their errors decreased so rapidly as to indicate that learning had taken place prior to that point. Blodgett's study showed, then, that learning may take place without a change in performance, implying that learning is a distinct process. Although it is true that learning is always inferred from performance, there may be principles unique to each process, and since it is possible to show that they operate independently (as in the Blodgett study) the distinction between them is a useful one.

Indices of Performance

Psychologists have devised techniques or conventions by which performance, and therefore learning, can be assessed. The major indices used to measure response strength are the rate or frequency of the response, the vigor, speed, probability, or the persistence of the response. Initially, the strength of a response is quite low if the subject has never had any experience with the response. However, as the subject learns, the probability of making the specific response increases over time. This increase in response strength measured by probability, speed, frequency, etc., represents learning.

Such an increase is called the acquisition phase of learning; the subject is acquiring the response. Acquisition, of course, does not occur accidentally; it is produced by some element in the training procedure. For example, a hungry rat will perform a response in order to be fed. However, if the food reward is not presented following a response, the index of performance will gradually decline until the subject finally stops responding, illustrating the extinction phase (see Figure 1-1 on next page). While speed, frequency, and other variables may be used to assess performance in both acquisition and extinction, the persistence of the behavior during extinction—the length of time required before the animal ceases to respond—can also be used to measure response strength.

Once a performance measure is chosen, it is then possible to vary certain variables in the situation to assess their effect on learning. For example, if a psychologist wishes to know the influence of a certain drug on learning, it is necessary to select an arbitrary response with which the subject has had no prior experience, administer the drug, and train the subject to perform the response. The effect of the drug on learning can be assessed by comparing the subject's performance with the performance of another subject that did not receive the drug.

Conditioning Models

In order to assess basic learning principles, arbitrary responses are studied and the experimental situation is simplified. In fact, the effort to study simplified situations has encouraged learning psychologists to isolate the basic units, or building blocks, of learned behavior just as

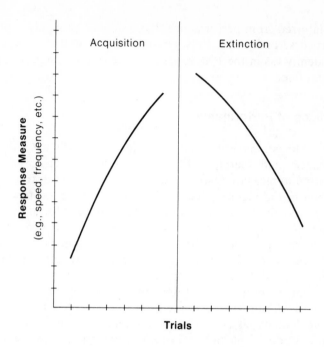

Figure 1–1. A hypothetical acquisition curve and extinction curve, plotted as a function of acquisition and extinction trials.

biologists have identified cells, chemists have utilized molecules, and physicists have referred to atoms as their basic units. Rather than a physical substance like molecules, these units are "learning units" or instances of simple learned behavior. Many psychologists feel that the basic units of learning, the building blocks out of which all complex learned behavior is fashioned, are the conditioning paradigms, or models, which specify a technique or method for training and are used to infer the nature of the underlying learning process.

There are two basic conditioning models, termed classical and instrumental. The remainder of this book is devoted to these two paradigms and the basic principles they reveal about the underlying learning process.

Summary

Most behavior is acquired or learned. While difficult to accurately define, learning occurs when, as a result of experience, a relatively permanent change in behavior can be observed. Changes in behavior due to maturation, fatigue, and other variables are not instances of learning.

Psychologists in the nineteenth century studied consciousness—mental ideas and sensations—by means of introspection. More recently, behaviorists, led by Watson, instead have emphasized that psychology must study overt behavior through the use of objective techniques. The study of learning has stemmed from this behaviorist movement.

Resulting learning research has been influenced by three major historical events or conditions. The first, the British Associationists, believed that mental associations developed from the contiguous occurrence of two sensations. The contiguity of two events was thought to be important in acquiring, or learning, an association. The second historical influence, Darwin's theory of evolution, emphasized adaptive behavior which was often achieved through learning. Finally, a unique social climate in America and Russia at the turn of the present century fostered the study of learning.

Learning psychologists today attempt to outline basic principles that apply to all learned behavior. To aid in such a complex task, they often study simplified responses in lower organisms such as the laboratory rat. From their observations, general principles of learning emerge and theories are formulated.

Through experimental investigations, many psychologists have found it useful to distinguish between learning and performance: Learning is an underlying process which determines overt behavior or performance. Learning is inferred from performance, although as shown in latent-learning studies, learning may be more extensive than the performance indicates. The techniques for measuring performance, and therefore for assessing learning, may include the probability, rate, or vigor of the response during the initial acquisition phase and, in addition, the persistence of the response during the extinction phase.

Classical Conditioning

Pavlov

Introduction

Classical conditioning is thought by many to represent a fundamental learning process as well as a technique for training a response. Classical conditioning has been demonstrated in a large number of different organisms, from humans to planaria, suggesting the procedures may, indeed, represent a fundamental learning process.

Appetitive Conditioning

Ivan Pavlov, a twentieth-century Russian physiologist, was engaged in the study of the physiology of digestion in dogs when he, almost accidentally, discovered classical conditioning. Pavlov noticed that when he entered the room in which the dogs were housed, they would begin to salivate before the experiment had even begun. He labeled these responses "psychic" secretions and thereafter diverted his efforts to the explicit study of these responses.

A basic example of Pavlov's classical appetitive conditioning experiment is as follows (for a more complete account of his work see Pavlov, 1927). Prior to the actual experiment, Pavlov had made a surgical incision in the cheek of the dog and attached a small glass tube to the salivary gland so that the amount of salivation could later be measured. After recovery from this operation and during the actual experiment, he first sounded a tuning fork for about 7 or 8 seconds (no salivation response). Immediately following the tone, he placed meat powder in the dog's mouth and observed that the dog immediately salivated. The combination of tone and meat powder was repeated approximately ten times after which the tone was given by itself for 30 seconds. Pavlov noticed a slight amount of salivation appearing in response to the tone only after about 18 seconds. After approximately 30 pairings, however, the dog salivated a great deal in response to the tone alone almost immediately after it was turned on. What the animal seemed to have learned was to anticipate the food powder. The tone, which originally was neutral, acquired the ability to elicit salivation.

As mentioned in Chapter 1, the Law of Contiguity stated that two events which occur together will become associated after repeated pairings. In this sense, Pavlov's experiment was an implicit test of the Law of Contiguity. In associating the tone with the meat powder, the dog became conditioned to salivate in response to the tone alone.

Defense Conditioning

Another important example of classical conditioning was first shown by Bekhterev (1913), whose method was essentially the same as in Pavlov's appetitive experiment. A neutral stimulus (tone) was sounded and was followed by an electric shock to the forepaw of the dog. Initially of course, the tone had no effect on the dog's behavior, whereas the shock always elicited a flex response. After repeated pairings, however, the response of flexing the forepaw appeared to the tone alone. This and other examples of such conditioning (e.g., Liddell, 1934) differ from Pavlov's experiment in at least one major way: The stimulus used to elicit the behavior initially was an aversive shock. In other respects, however, the defense conditioning experiment is analogous to the appetitive experiment.

Terms

The stimuli and responses which were used in these experiments are termed the following.

Unconditioned Stimulus

The unconditioned stimulus (US) is any potent or massive stimulus which evokes a regular and measurable unlearned response. In Pavlov's experiment the US was the food powder, whereas it was electric shock in Bekhterev's study. These stimuli have predictable effects on behavior, i.e., they always elicit a reflexive-type response over which the subject seems to have no control. For example, it would be difficult to prevent oneself from salivating if a drop of lemon juice were placed on the tongue. Likewise, a hot stove reliably, and virtually automatically, elicits withdrawal of the hand. Other US's commonly used in experiments are air puffs aimed at the eye (elicits blinking), mild shocks, thermal changes, etc.

Unconditioned Response

The unconditioned response (UR) is the regular and measurable unlearned response elicited by the US. In Pavlov's research it was salivation, while for Bekhterev it was the leg flex. Any response which is regularly elicited by a US is the UR, and salivation and leg flex are only two examples. The important aspect is the functional relationship between the US and UR, i.e., the UR is regularly elicited by the US. Many different UR's have been conditioned in classical experiments; examples are salivation, changes in galvanic skin resistance (GSR), pupillary reflex, nausea, muscle flex, blinking reflex, changes in respiration or heart rate, etc. Another important point is that many US's may elicit more than one UR at a time. This is most easily seen in the defense conditioning experiment, in which the shock elicits paw withdrawal but at the same time may elevate the heart rate and blood pressure, or alter the GSR.

Conditioned Stimulus

The conditioned stimulus (CS) is an originally "neutral" stimulus which precedes the presentation of the US. In Pavlov's experiment it was the vibrations of a tuning fork, although any stimulus may serve as a CS provided that it is perceived by the animal (a tone would obviously not function as a CS if the animal did not hear it). Examples of conditioned stimuli include various tactile, visual, and auditory stimuli as well as internal stimuli. The CS is not a potent stimulus as is the US. It is neutral in the sense that it has no dramatic effect on the organism as does the US, although the CS may evoke certain overt reactions such as head movements or eye closures if the organism perceives and attends to the stimulus.

Conditioned Response

The final term, conditioned response (CR), is the learned response which resembles the UR. In Pavlov's experiment the CR was salivation to the tone alone, while in Bekhterev's study, it was paw withdrawal in response to the tone alone.

It should be pointed out that a second response is learned in classical defense conditioning. Because aversive or noxious stimuli are used (e.g., shocks), the subject acquires a fear or aversion to the CS. In other words, the CS is paired with an unpleasant stimulus and, in addition to eliciting the motor flex, the CS comes to elicit fear as well. This point is highly significant not only because all defense experiments potentially involve fear or anxiety but also because classical defense conditioning may be a model (highly simplified) of phobias.

It could be argued that the CS in appetitive conditioning also comes to elicit an emotion in addition to the response of salivation. Such an emotion has been characterized by Mowrer (1960) as "hope," i.e., the CS signals a forthcoming pleasurable event (food). Mowrer's argument is logical although the emotion of fear (in the defense situation) often is more dramatically displayed than hope (in the appetitive case). (These points are discussed more fully in Chapter 4.)

Operations

The major operational characteristic of classical conditioning is that the events or stimuli are presented to the subject independent of the subject's behavior. That is, the response (either of salivating or flexing the paw) has no consequence on the presentation of the stimuli. The tone is always followed by the food powder or the electric shock; thus, the events are independent of the subject's behavior.

In effect, classical conditioning represents a technique for presenting stimuli during an experiment, as shown in Figure 2-1. Initially (in

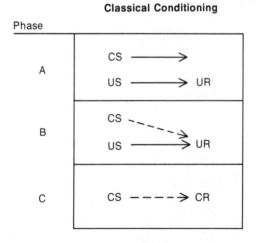

Classical Conditioning

Figure 2-1. Acquisition phases in classical conditioning.

phase A) the CS elicits no observable behavior, whereas the US elicits the UR. After repeated presentation of the CS and the US, an association develops between them as shown in phase B. Once conditioning has taken place (phase C), the CS is able to elicit the response (CR) in the absence of the US.

The preceding sections have referred to the establishment or acquisition of a CS-US association. On the other hand, after the acquisition phase the presentation of the CS alone (without following it by the US) weakens the response and produces extinction, i.e., the CS gradually loses its ability to elicit a CR.

Procedures

There are various ways of structuring the presentation of the CS and US in a classical experiment, all of which effect the type and strength of conditioning. The major procedures, shown in Figure 2-2, differ with respect to the temporal relationship between the CS presentation and the US presentation. The top line of the figure represents the US presentation; lower lines represent the CS presentation.

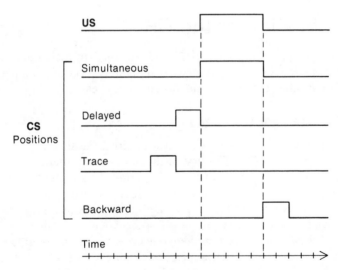

Figure 2–2. Four major classical conditioning procedures which differ with respect to the temporal relationship between the CS and US.

The first example is of simultaneous conditioning in which the CS is given at precisely the same time as the US. Simultaneous conditioning produces very little learning, since on subsequent test trials when the CS is presented alone, the CR is virtually nonexistent (Beecroft, 1966). The next line represents delayed conditioning. In this case, the CS is presented sometime prior to the US and continues at least until the onset of the US. The CS may go off, however, at various times either during or after the US. In contrast, in the trace conditioning procedure the CS comes on and then goes off *before* the US is presented. There is an empty interval between the CS offset and the US onset. In the last procedure, termed backward conditioning, the CS follows the US, and inferior learning takes place, although some weak learning effects have been shown. It is, of course, reasonable to assume that a dog will not salivate in response to a tone (anticipate food powder) if the tone follows the presentation of food powder. Numerous studies have shown that conditioning is superior in the delayed and trace situations relative to simultaneous and backward conditioning. That is, the CS must precede the US. Trace and delayed procedures are about equally effective in eyelid conditioning experiments (e.g., Ross & Ross, 1971) and some heart-rate studies (Wilson, 1969), although not in other heart-rate conditioning studies (Black, Carlson, & Solomon, 1962) which indicated that the trace procedure was less effective with longer empty intervals.

Specification of the CR

If learning principles are to be studied, it is extremely important to specify the CR accurately. Otherwise, there may be confusion between the CR and other erroneous, unlearned responses.

Measurement Techniques

There are basically two ways to structure a classical experiment in order to measure a CR. The first is the anticipation method, in which the CS is of a long enough duration to permit the subject to anticipate the US onset by responding before it occurs. There are problems, however, in using this technique. Different response systems vary according to their latency of onset; some are more "sluggish" than others. If the CS has to be lengthened in order to give more time for the response to occur, the optimum conditioning situation (CS-US contiguity) is worsened. For example, the GSR response has a latency of about 2 seconds (the time it takes for the response to appear following the CS); however, for GSR conditioning the optimum interval between the CS onset and US onset is only .5 seconds (White & Schlosberg, 1953).

The second method (test-trial technique) uses test trials interspersed throughout training, in which the CS alone is given for a longer duration than on training trials. There are problems with this method too; foremost is that on these test trials the US is omitted, thereby weakening the CR through extinction. That is, because conditioning strength generally increases with the number of CS-US pairings, US omission on the test trials reduces CR strength.

The fact that each technique has shortcomings simply means that the CR strength must be evaluated in light of these methods. Perhaps the best (or only) way to avoid the problems associated with these techniques is to pair the CS and US over several trials for different groups of subjects and then give only one test trial to each at the end of training.

Defining the True CR

An important point mentioned previously is that the CR only resembles the UR. It is usually somewhat weaker and slower to appear, and may contain only some components of the UR. For example, food powder may elicit both salivation and chewing responses, while the CS might elicit only salivation. In defense conditioning, the UR may include postural adjustments while the CR may be paw-flex exclusively.

Part of the problem of defining the true CR is that other responses (e.g., random, voluntary responses) also occur during conditioning. A classic defense experiment by Wickens (1939) illustrates the difference between a true CR and a voluntary response. A buzzer (CS) was paired with a mild shock (US) to the forefinger for one group. A second group was asked to make a voluntary finger-withdrawal response when a bogus US (a light) was turned on. The third group of subjects was similar to the second although they were threatened with shock (but never received it) for giving slow responses. Both of the voluntary groups (groups 2 and 3) showed very little conditioning; i.e., their responding was slow and variable when given the buzzer alone. The first group, however, gave many responses to the buzzer (CS) and continued to do so after the shock was discontinued. Wickens concluded that true conditioned CR's differ from voluntary responses in terms of their latency and persistence.

The problem of specifying the CR goes considerably beyond the issue of voluntary responses. Many investigators have noted that responses are given, at times, without the forward conditioning procedure (delayed or trace) being in effect. Such responses are called pseudoconditioned and are not true CR's. Pseudoconditioning is defined as an inflation in the CR strength, usually temporary, that is attributable to factors other than CS-US pairings. Although the factors responsible are not well known, it appears that extraneous cues or motivational variables can produce pseudoconditioned responses. According to Kimble (1961), pseudoconditioning may be present in all forms of defense conditioning (e.g., where a noxious US is used), suggesting the involvement of emotionality.

Control Procedures

The way to deal with pseudoconditioning, in fact the way to identify the characteristics of the CR, is by the proper use of control procedures. Conditioning in the experimental subjects may be compared to the level of responding in control groups. A control procedure is one that provides the control subjects with essentially the same conditions as those given to the experimental subjects except for the one important feature that produces conditioning (contiguous CS-US presentation).

Control procedures have been discussed by Beecroft (1966) and elaborated upon in a theoretical paper by Rescorla (1967b). The basic controls for pseudoconditioning are: (a) to present the CS only; (b) to present the US only; (c) to present both the CS and US but randomize their presentation such that they are not explicitly paired; (d) to present the CS following the US (backward conditioning); and (e) to present the CS at the same time as the US (simultaneous conditioning). In each of these procedures, the subject receives one or both of the stimuli, but the essential feature in classical conditioning (CS-US contiguity) is not present. Presumably, then, each of these procedures would yield very little, if any, true conditioning.

The first two control procedures are clearly deficient to the extent that the control subjects do not have the same experience as the experimental subjects. In the first case, the groups differ in terms of US experience, and in the second case, in terms of CS experience.

The last three procedures are better in this regard: Both groups of subjects receive both stimuli although they are paired in a forward manner for only the experimental group. Backward conditioning would seem the least desirable of these three procedures since CR's have been found using this procedure. For example, Spooner and Kellogg (1947) found that backward conditioning did lead to a conditioned finger-withdrawal, although the strength of the response was very low in contrast to the forward-conditioning groups. Finger withdrawal in response to the CS was given on an average of only about 10 percent of the test trials.

Rescorla (1967b) has argued that the random procedure is the best. In this method, the control subjects receive both CS's and US's but in a random, independent fashion; the occurrence of the CS provides essentially no information to the subject about the occurrence of the

US and thus no association develops. The method prevents the CS from acquiring the ability to signal to the subject not only the occurrence of a US but also the nonoccurrence of a US (which is the case in backward conditioning). The CS is simply irrelevant to the US. Despite Rescorla's claim for the appropriateness of this control procedure, the picture is not entirely clear. Some recent studies have suggested that with the random procedure, the CS does develop an ability to elicit a CR (Benedict & Ayres, 1972; Kremer, 1971; Kremer & Kamin, 1971; Quinsey, 1971).

It is possible that simultaneous CS-US presentation is the best control since evidence suggests it produces the least conditioning (Beecroft, 1966). One problem is that it is not always possible to control the stimuli onset accurately enough to insure simultaneity of presentation. Although several experiments have compared most of the procedures listed above (Prokasy, Hall, & Fawcet, 1962) few have included all of the procedures in the same design.

Inhibition

Up to this point, only excitatory conditioning has been discussed, i.e., conditioning in which the CS excites or elicits the response. However, inhibitory conditioning may also take place, in which the CS will prevent or inhibit the occurrence of the response. In the discussion on control procedures, it was stated that the random control method prevents the CS from acquiring the ability to signal either the occurrence or nonoccurrence of the US. What was implied, of course, was that the CS can give information about the nonoccurrence of the US if the CS is explicitly paired with that event. When this happens, the CS does not excite or elicit the CR but rather, it becomes inhibitory and suppresses the CR.

Extinction and Spontaneous Recovery

During extinction, the US is never presented, i.e., the CS occurs alone. This is one example of the CS being explicitly paired with the nonoccurrence of the US. The CS *is* providing information about the nonoccurrence of the US just as it provided information about US presentation during acquisition. Consequently, the CS becomes inhibitory and suppresses the CR.

The decline in CR strength during extinction traditionally has been thought to represent exactly this fact: Extinction involves an active inhibitory process which counteracts the excitatory effects. The major evidence for this is that the effects of extinction are not permanent. In fact, the CR reappears (at least partially so) if the subject is given a rest period following extinction. This phenomenon is called spontaneous recovery, examples of which were first reported by Pavlov (1929). For instance, he observed in one demonstration that the dog salivated 10 drops of fluid to the sound of the CS at the start of extinction. At the end of extinction the response was down to 3 drops, while after a 23-minute rest period, the response increased again to 6 drops. Other examples of spontaneous recovery of an appetitive response (Lewis, 1956) or a defense response (Riess, 1971) support the notion that extinction, in part, involves a temporary suppression of the CR. However, some recent work (Rescorla, 1967a) has, in fact, questioned whether extinguished CS's are always inhibitory.

Conditioned Inhibition

In more general terms, conditioned inhibition refers to the acquisition of inhibitory properties by any CS. An inhibitory CS is not neutral nor does it simply distract the subject; rather, it produces a tendency to not respond (for a review, see Rescorla, 1969).

There are several ways to measure conditioned inhibition. One may be through spontaneous recovery as mentioned above. However, a clearer measure of CS inhibitory properties is by a technique called summation. During conditioning one stimulus is paired with the US and it becomes excitatory (CS+) while a second stimulus is paired with the nonoccurrence of the US and it becomes inhibitory (CS−). On a test trial, then, both the CS+ and CS− are given together, and due to the counteracting tendencies (excitatory and inhibitory), a smaller CR is observed than when the CS+ is given alone. Thus, the addition of the inhibitory CS− reduces the CR strength.

Another basic way to measure conditioned inhibition is to establish a CS as inhibitory prior to its becoming an excitatory CS+ and observe that acquisition of the excitatory effects is retarded. An excellent example of this is found in the classic defense experiment by Lubow (1965), in which goats and sheep were first presented with a circular patch of light (CS) for either 0, 20, or 40 trials. No US was given at this time. During the subsequent acquisition, however, a shock (US) was presented to the foreleg following the CS, and the flex (CR) was measured. The CS, of course, developed inhibitory tendencies during the pre-exposure trials and subsequently inhibited acquisition of the leg flex. As shown in Figure 2-3, the percentage of CR's given on the four acquisition days was lowest for the group that had previously received 40 CS pre-exposures (strongest conditioned inhibition) and highest for the group that received no CS pre-exposures (no conditioned inhibition). Thus it is clear that stimuli acquire excitatory properties if followed by a US, but inhibitory properties if not.

Inhibition of Delay

There is one instance in which a CS may be followed by the US and yet (in contrast to the principle stated immediately above) develop inhibitory properties. This occurs when the CS-US interval is reasonably long (often as long as 30 seconds). Over training, the subjects gradually begin to inhibit or suppress the response at the onset of the CS as is usual. Instead, the response is given closer and closer to the scheduled US. For example, Kimmel (1965) trained subjects in a GSR experiment using a red light (CS) which stayed on for 7.5 seconds and terminated with the shock (US). Fifty trials were administered and the latency of the GSR response was measured. As shown in Figure 2-4 (see next page), subjects took longer to respond with greater training.

How is it known that an inhibitory process was at work in Kimmel's experiment? This question was answered by showing that the response latency was restored to its original time if the subject was

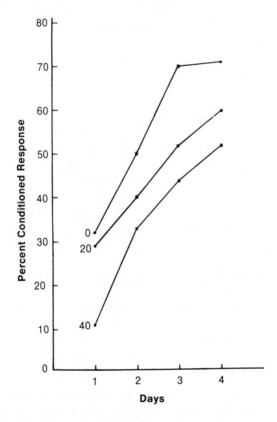

Figure 2-3. Percentage of conditioned responses on four acquisition days as a function of number of CS preexposures.

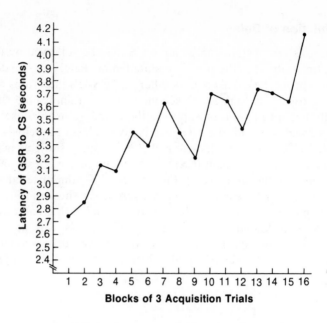

Figure 2–4. Mean conditioned response (GSR) latency as a function of blocks of three acquisition trials.

disrupted (during the inhibition) by a novel stimulus. This phenomenon of disruption is called disinhibition. In Kimmel's study, a novel tone, sounded during the CS, was the disinhibitor. Although the response took about 4.2 seconds to appear to the CS after the 50 conditioning trials, it dropped to 2.35 seconds on 4 subsequent disinhibition trials. That is, when appropriately distracted by a novel cue, the subjects lose the inhibitory tendency which is suppressing their behavior at that moment and they respond immediately.

Factors in Classical Conditioning

Knowing that a response may be classically conditioned, it has been the task of many psychologists to investigate systematically the variables affecting conditioning.

CS-US Interval

The CS-US interval is an important facet of a classical experiment and a variable that has received a great deal of attention. One reason for its importance is that the interval essentially defines CS-US contiguity and therefore the limits of classical conditioning. Generally, as the CS-US interval is lengthened, the contiguity relationship is weakened and conditioning decreases.

There is some dispute about the optimum CS-US interval. Ross and Ross (1971) confirmed some earlier work in showing that .5 second is optimum for eyelid conditioning. Beecroft (1966) claimed that .5 second is also optimum for finger withdrawal and GSR in humans.

This optimum interval varies, however, for other responses and for different species of animals. For example, eyelid conditioning in rabbits was inversely related to the CS-US interval with the best conditioning occurring at .2 second for Schneiderman and Gormezano (1964) or .25 second for Smith (1968). In contrast, heart-rate conditioning was best with an interval of 2.5 to 10.0 seconds in dogs (Black, Carlson, & Solomon, 1962), 2 to 6 seconds in rabbits (Deane, 1965), and 13 seconds in humans (Hastings & Obrist, 1967). The optimum interval is certainly always short, i.e., measured in seconds rather than in minutes, although it does vary with the type of response and species, and numerous exceptions to the .5-second interval exist.

Intertrial Interval

The distribution of trials has also been shown to affect conditioning. Most of the work has been done in eyelid conditioning, where it has been shown that short intertrial intervals produce inferior conditioning (Spence & Norris, 1950). This effect has been more recently supported by Prokasy and Whaley (1963) where intervals of 20 to 35 seconds gave better performance than shorter intervals of 5 to 20 seconds. Variability in the intertrial interval, however, seems to have little effect (Prokasy & Whaley, 1961; Prokasy & Chambliss, 1960).

US Intensity

A great deal of work has been done relating US intensity to the strength of the CR. Generally findings indicate that CR strength increases as a function of US intensity up to some specifiable limit. For example, Trapold and Spence (1960) showed that eyelid conditioning in humans was stronger when the air puff to the cornea was 2.0 psi. as opposed to .25 psi. Beck (1963) confirmed this result using .5 versus 5 psi. as the US intensity, and analogous results have been shown more recently for eyelid conditioning in rabbits. Smith (1968) used shocks ranging from 1 to 4 mA in intensity and found that the CR strength (percentage of CR's on the test trials, and amplitude of membrane closure) increased with US intensity.

When considering heart-rate conditioning in rats, several recent studies have shown better conditioning with higher US intensities

(Caul, Miller, & Banks, 1970; Fitzgerald & Teyler, 1970). In the latter experiment, shock ranged from .4 to 5.0 mA in intensity. The mean heart-rate deceleration is shown in Figure 2-5. Both the delayed and trace conditioning groups improved with higher US intensities (the delayed being superior to the trace) although the delayed groups' performance decreased with further increases in US intensity beyond 1.2 mA. It appears therefore that the general relationship between performance and US intensity holds for a variety of species and responses.

US Duration

Several studies have shown that a longer US duration does not have the same effect as a more intense US. For example, Wenger and Zeaman (1958) compared shock durations of .1, .2, .6, or 1.5 seconds in human heart-rate conditioning and found CR strength unrelated to duration. Overmier (1966) confirmed this finding, but also showed that the effect of US duration upon conditioning depends on the measure

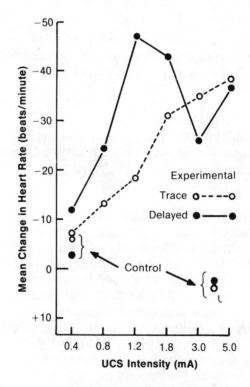

Figure 2–5. Mean change in heart rate, using both trace and delayed conditioning procedures, as a function of US intensity.

employed. While the extent of heart-rate conditioning was not influenced by US duration, motor conditioning was found to be stronger with a 50-second US than a .5-second US.

CS Intensity

Contradictory results have been obtained concerning the way CS intensity affects classical conditioning. Furthermore, it is uncertain whether CS intensity affects learning (i.e., strength of the association) or whether it simply modifies performance.

Although in eyelid experiments a change in conditioning as a function of CS intensity was not found by several investigators (e.g., Grant & Schneider, 1948), other experiments have shown positive results (e.g., Beck, 1963; Moore, 1964). For example, Beck trained subjects with two distinct CS's (a 30 and 80 db tone) and found superior conditioning with the stronger CS. In investigating the eyelid reaction in rabbits, Gormezano (1972) has recently shown that higher levels of responding are obtained with more intense CS's. He found that the effect appeared not only for two groups differing on the basis of CS intensity but also (as in Beck's study) within a single group receiving both high and low CS intensities. In summary, it appears that a more intense CS very often elicits a stronger CR, but the exact conditions which determine this effect are poorly understood (see Champion, 1962, for a review).

Instructions

It was stated previously that true CR's differed from voluntary responses. Indeed, most classically conditioned responses (e.g., heart rate, GSR, salivation) are nonvoluntary and "automatic." It is curious that verbal instructions would modify conditioning, but a great deal of evidence indicates that they do. However, the effects are in one direction only: It is clear that negative verbal instructions inhibit or reduce the extent of conditioning, while positive instructions simply have the effect of eliciting more voluntary responses as opposed to true CR's.

A study by Norris and Grant (1948) illustrates the negative effect of instructions. They administered 75 eyelid conditioning trials to groups of subjects; half were warned not to blink before they felt the air puff (one group was even given a mild wrist shock as punishment for doing so), while the others did not receive instructions. The results clearly demonstrated that the negative instructions depressed CR performance although not completely so.

In contrast, positive instructions (telling the subjects to blink so as to avoid the air puff) do not strengthen the CR. Gormezano and Moore

(1962) investigated the effect of such instructions on eyelid conditioning. They found that under these instructions, subjects gave a higher percentage of responses, which were of shorter latency, but the responses differed markedly from those in the noninstructed group. Namely, they were very sharp closures, characteristic of voluntary responses. Thus the improvement in performance was not in terms of better conditioning but rather in terms of converting the CR's to voluntary responses.

Effect of Omitting the US Presentation

When the test-trial technique is used to measure a CR, the US is not presented on that trial, i.e., the CS occurs by itself. This was cited earlier as a problem with this technique insofar as the CS-US association would be weakened. However, the effect of explicitly omitting the US has been studied.

The general finding is that US omission does lead to a lower level of conditioning (and thus the admonition regarding this technique is justified). Moore and Gormezano (1963), in studying eyelid conditioning, presented the US on either 25, 50, or 75 percent of the 80 acquisition trials to separate groups. They found that the mean percentage CR's given by the 25-percent group was 56.6 percent whereas this figure was 78.2 percent and 83.4 percent for the 50- and 75-percent groups respectively. Thus, the conditioning was better with a higher percentage of US presentations. In addition to this finding, Moore and Gormezano noted that the US-omission technique provided higher levels of conditioning than if the US's were merely delayed, strongly suggesting that the test-trial technique, despite the significant weakening of the association by US omission, is superior to the anticipation technique.

Special Cases of Conditioning

So far, only conventional responses and stimuli have been considered. However, there are several significant examples of learned behavior which appear to be conditioned classically but which deviate somewhat from the salivation or GSR paradigms. Four of these are considered below.

Sensory Preconditioning

In a sensory preconditioning experiment, two neutral stimuli are first paired, one is then used as a CS in a normal conditioning experiment, and finally the other is substituted as the CS. Usually it is observed that the substituted CS is effective in eliciting the CR even though it was not actually used in the conditioning phase. In an example of sensory preconditioning of GSR, Coppock's (1958) preconditioning group (PC) received 10 tone-light pairings (the tone lasted 3 seconds with the light coming on for the last 2 seconds). The control group (C), on the other hand, received 10 preconditioning trials but the light and tone were not paired as they were in the PC group. A third group (IPC) was given light-tone preconditioning trials—the stimuli were in the reverse order from the PC group. For all groups, the light was then used as the CS in conditioning the GSR response followed by substitution of the tone for 5 trials. The results, shown in Figure 2-6, indicate that virtually no responses were elicited by the tone CS in 5 test trials in groups C and IPC; thus, the tone CS was virtually neutral. However, in group PC substantial responding was found, and the tone appeared to have excitatory properties. The point of the study is that by simply pairing two stimuli an association seemed to develop. Lack of pairing (as in group C) or backward pairing (as in IPC) did not produce the same effect.

Is the initial preconditioning phase an example of classical conditioning? Earlier it was suggested that classical conditioning requires a US, a stimulus that produces a large, reliable, reflexivelike response. This principle seems to be violated by the sensory preconditioning studies. How can sensory preconditioning be considered an example of classical conditioning if it does not involve a suitable US and the corresponding UR?

One resolution of this paradox is to suggest that the sensory preconditioning stimuli actually do elicit UR's (e.g., orienting or

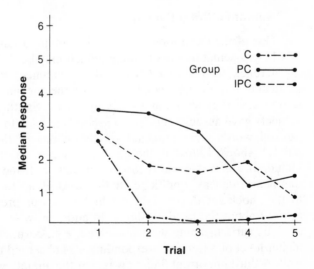

Figure 2–6. Median GSR response as a function of test trials.

attending responses) but that they are small and difficult to observe or measure. The phenomenon might then operate as shown in Figure 2-7. During the preconditioning phase, R_2 is actually functioning like a UR (to which CS_1 becomes associated). The response R_1 is an intermediate or mediating step in this association. In the normal conditioning phase, CS_2 comes to elicit the CR. Finally, on the test, the CS_1 also elicits the CR by first eliciting R_2 (which had been conditioned during the preconditioning phrase). If such were true, sensory preconditioning is like a double experiment with two separate associations later combined on the test trials. During the test, the first association (R_1 established during preconditioning) is elicited, which in turn elicits the second association (the CR which was established during normal conditioning). Therefore, sensory preconditioning suggests that classical conditioning, in general, is not limited to massive responses (e.g., UR's) but rather involves associations between any stimuli regardless of their intensity or biological significance. The necessity of even weak UR's in establishing associations, however, is questionable since Cousins, Zamble, Tait, and Suboski (1971) have demonstrated sensory preconditioning in curarized rats (totally paralyzed by injections of curare and unable to respond). Regardless of the way in which sensory preconditioning works, these experiments appear to illustrate the minimal conditions necessary for association learning, i.e., contiguity of two stimuli without regard to the nature of the elicited responses (see Seidel, 1959, and Thompson, 1972, for reviews of sensory preconditioning).

Semantic Conditioning

One of the three topics reviewed in an important paper by Razran (1961) was semantic conditioning, which may be defined as conditioning a reflex to a verbal CS (word or sentence) irrespective of the particular letters, i.e., conditioning to the meaning of the word or sentence. For example, in a study by Lacey, Smith, and Green (1955), subjects gave associations for 15 seconds to each to 40 words on a list. Several words were repeated six times, one of them being *cow*. A painful shock followed the last association to these words (cow included). Heart rate was recorded for each trial, and the authors found that heart rate was conditioned to the word. Heart rate was higher prior to the shock for those trials in which cow was presented, indicating anxiety about the pending shock. In addition, when other words with obvious rural meaning were presented (e.g., corn, chicken, tractor), a higher level of autonomic responding was observed to those stimuli as well. A third important finding was that the anxiety which the subjects developed about the shock was unconscious in the sense that the

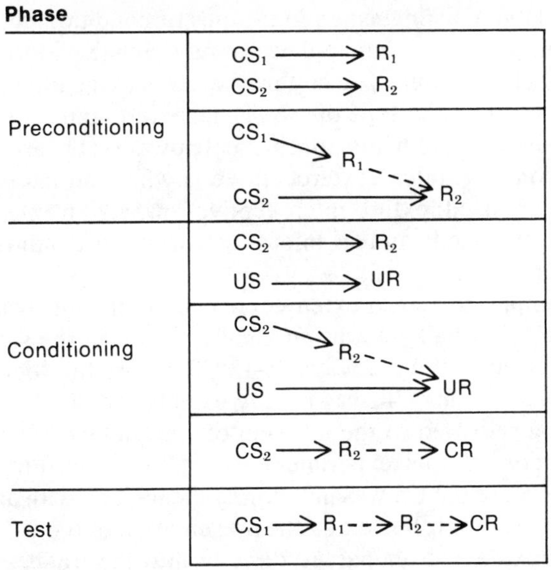

Phase

Preconditioning	$CS_1 \longrightarrow R_1$ $CS_2 \longrightarrow R_2$ $CS_1 \searrow$ $\quad R_1 \dashrightarrow$ $CS_2 \longrightarrow R_2$
Conditioning	$CS_2 \longrightarrow R_2$ $US \longrightarrow UR$ $CS_2 \searrow$ $\quad R_2 \dashrightarrow$ $US \longrightarrow UR$ $CS_2 \longrightarrow R_2 \dashrightarrow CR$
Test	$CS_1 \longrightarrow R_1 \dashrightarrow R_2 \dashrightarrow CR$

Figure 2-7. The way in which implicit responses (R_1 and R_2) may mediate the sensory preconditioning experiment.

subjects never knew when shocks were to be delivered. In summary, subjects demonstrated classical conditioning to a word CS, and these effects transferred to other stimuli that resembled the CS word only in general meaning or connotation. Thirdly, the subjects seemed unaware of the word-shock relationship.

This phenomenon is of enormous potential significance, since it provides a framework for considering some unconscious emotional processes. It also illustrates that conditioning may occur with stimuli quite unlike those mentioned previously (e.g., tones and lights) and that the conditioning may occur unconsciously. One implication is that much of man's conceptual or intellectual life may be fundamentally related to conditioning. Words and symbols take on meaning by virtue of the stimuli with which they are associated. Semantic conditioning may provide a model for investigating the development of such complex behavior on a simple level.

Interoceptive Conditioning

A third type of classical conditioning which is also somewhat unique is interoceptive conditioning. It is defined as the situation in which either the CS and/or the US are internal, i.e., applied directly to the visceral organs or other internal parts of the body.

Razran (1961) distinguished four types of conditioning. The prefix intero refers to internally applied stimuli, while extero refers to externally applied stimuli. The first prefix refers to the type of CS while the second refers to the type of US. Thus, extero-exteroceptive is the typical paradigm where lights, tones, meat powder, etc., are used as the stimuli. In contrast, intero-exteroceptive is where an internal CS and external US are used; extero-interoceptive is where an external CS but internal US are used; and in intero-interoceptive conditioning both stimuli are applied internally.

An example of intero-exteroceptive conditioning was cited by Razran in which a balloon was surgically placed in the stomach of a dog and irrigated with cold water (8-12°C.) before the food (US) was administered. A strong CR was readily established, and, within 6 to 9 trials, the dog salivated to the infusion of water alone.

A variation of this experiment exemplifies intero-interoceptive conditioning. Here the CS was inflation of the balloon with air (to cause stomach distension), followed by the presentation of the US (a mixture of carbon dioxide and air passed directly into the trachea through a surgically implanted tube). The gaseous mixture caused a defensive respiratory response. After only 3 to 6 pairings of these stimuli, the dogs began to demonstrate learning; they gave the defensive respiratory response to the stomach distension only.

A final example is of extero-interoceptive conditioning. Humans, who for medical reasons had balloons surgically implanted in the bladder, observed an arrow on a dial move (CS) while the balloon was inflated (US). Inflation produced the urge to urinate (UR). After several pairings when the dial was moved (by the experimenter) but the balloon was not inflated, the subjects reported the urge to urinate even though no physical cause (bladder distension) was present.

It is clear, particularly from the last example, that interoceptive conditioning is highly significant for understanding some instances of complex behavior. Interoceptive conditioning may lead to unconscious reactions—man's unconscious feelings may depend, in part, on simpler conditioning phenomena. Furthermore, interoceptive conditioning provides a model for dealing with psychosomatic illness; physical complaints may develop (without the person being aware) that have their basis in classical conditioning. Lastly, interoceptive conditioning illustrates the generality of classical conditioning to a wider variety of stimuli and responses.

Acquired Taste Aversion

The final example of learning which is significant yet different from normal classical conditioning experiments involves the acquisition of an aversion to a flavored substance. It is clear that aversions to or fears of a CS may be learned in a classical defense experiment, but in this case, subjects acquire an aversion to a flavored substance (CS) when the US (e.g., X-irridation, lithium, or other substances which cause illness and nausea) is presented after minutes or hours rather than after only seconds as in typical studies. This phenomenon seems to contradict the principle of contiguity, in that the CS-US interval seems to be far too long to support classical conditioning.

An example may be found in the study by Kalat and Rozin (1971), in which rats, deprived of water for 24 hours, were given a 10 percent sucrose and water solution (CS) in their cage for 2.5 minutes, followed either .5, 1, 1.5, 3, 6, or 24 hours later by a 6-ml. injection of lithium chloride (US). Two days later (also while thirsty) the subjects were presented with the sucrose solution and plain water. The measure of aversion was the amount of sucrose drunk relative to the total fluid intake (normally rats will drink sucrose water almost exclusively and ignore plain water). The results, shown in Figure 2-8, indicate that when the pairing was delayed for 24 hours, little aversion for sucrose was shown, and no association developed between the solution and the illness. However, for shorter delays, strong aversion was established so that for the group given lithium chloride 30 minutes after the CS solution, only about 5 percent of its total fluid intake was sucrose.

The results resemble classical conditioning in the sense that poor conditioning was observed when the CS and US were further sepa-

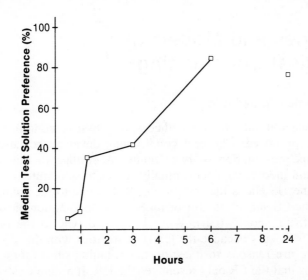

Figure 2–8. Median percent intake (relative preference) of sucrose solution as a function of the solution-poison delay during training.

rated in time. However, the fact that learning was achieved with delays of hours, rather than minutes, suggests that acquired taste aversion is outside the scope of normal conditioning (see Garcia, McGowan, & Green, 1972 and Rozin & Kalat, 1971 for reviews). There are other anomalies associated with the taste aversion phenomenon and some are discussed in Chapter 10. For the present, the point is simply that additional principles may be required to account for this significant finding although, superficially, acquired taste aversion comes under the heading of classical conditioning.

In summary, the preceding examples of learning are similar to classical conditioning yet they deviate in many respects. In fact, principles in addition to those already mentioned may be needed to account for these phenomena. However, each phenomenon represents an important extension or implication of classical conditioning for complex behavior outside of the laboratory. Thus, sensory preconditioning illustrates association of ordinary events, stimuli not typically used in the laboratory. Semantic conditioning illustrates a possible basis for the development of unconscious, conceptual (verbal) behavior. Interoceptive conditioning suggests a model for unconscious bodily feelings, and a basis for psychosomatic illness. Finally, acquired taste aversion implies a basis for the bait-shy behavior of some animals. Thus classical conditioning is a useful model of learning, and by studying the factors which influence conditioning in simple situations, knowledge may be gained about other important and complex behaviors.

Theories and Models of Classical Conditioning

Stimulus Substitution Theory

Throughout the chapter a theory of classical conditioning was implicitly proposed: Classical conditioning develops because of the contiguous presentation of two stimuli. According to this stimulus substitution theory, the CS eventually comes to substitute for the US; initially the US elicits the response (UR) but due to the contiguity of stimuli, the CS substitutes for or takes over the function of the US in eliciting the response (CR). What is learned (CR to the originally neutral CS) occurs because the CS is contiguous with the US.

There are reasons to question the stimulus substitution theory. The first is that the CR only resembles the UR. If a simple substitution was achieved, it is logical that the two responses should be identical. However, the learned response is different from the unlearned response. A second criticism is that there are some reflexes (e.g., pupil dilation and contraction for which the US is a light) that simply cannot be conditioned (Young, 1954). This is a problem for the stimulus substitution theory because a UR is certainly observed to the light and yet the CS does not come to substitute for the US in eliciting that response. A third problem for the theory is that observable uncondi-tioned responses, in fact responses in general, are not required. This appears true from Cousins' et al. (1971) experiment where sensory preconditioning took place in paralyzed subjects, as well as from other experiments showing classical conditioning in curarized subjects (e.g., Solomon & Turner, 1962).

One solution, based largely upon this third criticism, is to ac-knowledge contiguity as important in conditioning but disclaim that classical conditioning involves or requires either the use of biologically significant stimuli, or overt responses from the subject. Rather, what appears to be more important (according to Rescorla, 1967b) is the formation of an association between a stimulus and its consequence or outcome. Stimuli (CS's) acquire the capacity to inform the subject about a future consequence (US), and the association develops not simply as a function of CS-US contiguity but rather as a function of CS-US correlation. A stimulus explicitly not followed by a US does not merely fail to acquire an association, it develops inhibitory properties. Similarly, CS-US contiguity on 50 percent of the trials doesn't produce half the associations that are produced with continuous pairing, but rather a combination of excitatory and inhibitory tendencies. In summary, Rescorla's theory of classical conditioning states that CS-outcome contiguity (or correlation) is the principle involved in the learning of associations. This differs from the original stimulus substi-tution theory which claimed that CS-US contiguity alone (where the US is a potent stimulus) is the principle involved in learning.

Modified Conditioning Theory

Rescorla's theory of classical conditioning stated above has been expanded (e.g., Kamin, 1969; Rescorla & Wagner, 1972; Wagner, 1969) to include multiple-cue learning. In most learning situations there are many cues which may act as CS's. The Law of Contiguity predicts that an equal association would develop to all cues—all of the stimuli are correlated to the outcome and therefore each cue would acquire the same strength to signal or inform the subject of the outcome. A great deal of evidence now indicates that this is not true. Associations develop selectively to cues depending upon the configuration of the cues.

An example is a study (Wagner, 1969) in which rabbits were divided into three groups in an eyelid-conditioning experiment. First, a compound CS was presented (light flashes plus a tone) followed by a mild shock. One hundred twelve such trials were given to group II (the control group). For group I, interspersed with these trials, there were again as many trials where the light flashes were presented alone and followed by shock. Group III also got additional light-only trials, but the light was not followed by shock. All groups were subsequently given 16 test trials, this time with the tone (rather than the light). In summary, the experiment concerned the effect of additional shock (or nonshock) trials for one of the elements in the original CS compound (the light) on the associative strength of the other element (the tone). Would a change in the associative strength of the light affect the associative strength of the tone? Traditional contiguity theory would predict that since all the groups had the same number of tone-shock pairings, the associative strength of the tone should not vary between groups.

The results are shown in Figure 2-9. Associative strength of the tone was reduced in group I (relative to the control group II) while it was elevated in group III. That is, by increasing the associative strength of the light (group I), the associative strength of the tone was reduced. On the other hand, reducing the strength of the light (by additional lights without shock as in group III) increased the associative strength of the tone.

It appears from these data that stimuli in a compound compete for

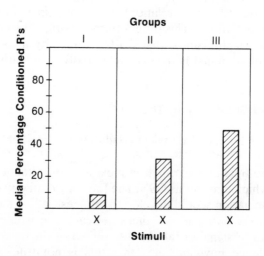

Figure 2–9. Median percentage conditioned eyeblink response to the tone for the three training groups.

strength. If one stimulus is strong (through extra conditioning as in group I) the other one becomes weak in associative strength. In contrast, if one stimulus is weak (through extra trials without the shock US, as in group III above) the other becomes relatively stronger. The important principle here is that associative strength depends not only upon stimulus contiguity but also on the momentary associative strength of other stimuli in the situation.

Another important principle in the new theory of classical conditioning was demonstrated by Kamin (1969). Kamin established fear in response to a CS by pairing it with a noxious US. He then presented the CS in another situation where the animal pressed a lever for food. The more the animal's behavior was suppressed by the fear-evoking CS, the greater the fear. In other words, degree of disruption of the ongoing behavior by a previously conditioned CS was used as a measure of associative strength to that CS.

Of Kamin's two basic groups, group C (control) received 8 trials of classical conditioning where a compound CS (light and tone) was followed by a shock (US). The light alone was then presented to the subject (while the subject was lever-pressing) and Kamin observed that it produced nearly total suppression of the rat's lever-pressing behavior, i.e., the light alone elicited a strong fear reaction. Group B, on the other hand, had received 16 tone-US trials (no light) prior to the 8 compound conditioning trials. Fear of the light was then tested in the same manner. (Note that both groups had the same number of light-shock pairings during the 8 compound CS conditioning trials.) Virtually no fear of the light was displayed by group B. The groups were treated identically except that group B had prior tone-only trials. Both groups were tested for associative strength to the light only. Prior conditioning trials to the tone (group B) blocked or prevented conditioning to the light. As in the Wagner experiment, the associative strength to the tone was high and therefore blocked the light from becoming an effective CS when given in the compound.

This extraordinary finding suggests that a subject has to be surprised, i.e., pay attention in order to learn. If an element in the CS compound already predicted the US (as the tone did for group B) the addition of the light in the compound added no new information about the shock US; it was redundant. Additional experiments helped confirm that conditioning to the light takes place only if the subject is surprised, i.e., if the added CS element gives new or further information about the US. This new principle—associative strength to stimulus is produced when it is informative or surprising but not when redundant—is of great importance, for it illustrates that while CS-US contiguity is a crucial underlying principle, it is not a sufficient explanation of conditioning.

Summary

Classical conditioning, as first discovered by Pavlov, represents a fundamental learning process as well as a procedure for training responses. In a classical conditioning experiment, the unconditioned stimulus (US) is the potent stimulus which regularly elicits the unconditioned response (UR). The conditioned stimulus (CS) precedes the US and later elicits the learned, conditioned response (CR) by itself.

The important feature in classical conditioning is that the CS and US are paired independently of the subject's behavior, although the temporal position of the CS with respect to the US may vary. The conditioned response, elicited in anticipation of the US or on a test trial, differs from voluntary or pseudoconditioned responses which are not dependent upon contiguous CS-US presentations.

In addition to the excitation of a response, a CS may become inhibitory and suppress a CR. Extinction represents an inhibition process since, if followed by a rest interval, the CR will spontaneously recover. A CS may also become inhibitory if it is consistently paired with the nonoccurrence of the US or with a delayed US.

One of the most important variables which affects the strength of a CR is the CS-US interval. While .5 seconds is optimum for many responses, the optimum interval may vary from .2 to 10.0 seconds. Conditioning is weaker if the intertrial interval is short or if the US is relatively weak. Conflicting results have been obtained regarding CS intensity, although several studies show that stronger CS's lead to better conditioning. Finally, verbal instructions may reduce conditioning, as will the omission of the US.

There are several unique cases of classical conditioning. The first, sensory preconditioning, doesn't use a potent biological US. Rather, these experiments indicate that an association may develop between any two relatively neutral stimuli. In the second, semantic conditioning, words or pictures are used as CS's, while the third case is interoceptive conditioning in which the CS and/or US is an internal stimulus. The fourth unique situation—acquired taste aversion— involves a subject given a flavored substance followed by lithium chloride which produces nausea; an aversion develops to this substance even though the poison occurred hours later.

Traditionally the theory of classical conditioning postulated that the CS comes to substitute for the US due to their contiguity. However, more recent experimental work indicates that the correlation between the CS and US is important in determining the strength of conditioning. This new interpretation has been applied to multiple-CS learning. If one stimulus is used as a CS, and a second is added to make a compound CS, the excitatory strength of the first will block conditioning to the second, suggesting that stimuli compete for strength.

Instrumental Conditioning

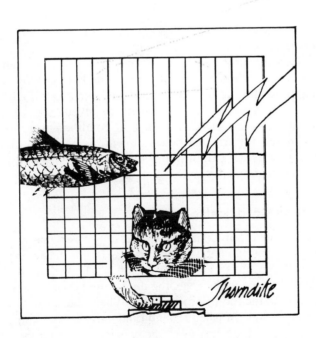

Thorndike

Introduction

It was suggested in Chapter 2 that classical conditioning represented one fundamental learning process. Many psychologists believe there is a second learning process, instrumental (or operant) conditioning, which differs from classical conditioning in its laws and fundamental underlying principles. This issue, whether one or two unique learning processes exist, has certainly not been resolved (see Chapter 4). However, at the very least, instrumental conditioning represents a unique technique for training behavior and, as such, is a focus for establishing functional relationships between environmental variables and the behavior they effect.

Appetitive Conditioning

E. L. Thorndike, of Columbia University, was a major innovator in the field of instrumental conditioning. In his basic experiment (Thorndike, 1898), hungry cats were placed in a locked puzzle box. By hitting a latch located in the cage, the door would spring open allowing the cats access to the food placed nearby. When Thorndike first put the cats in the cage, they thrashed around randomly as if seeking a way to escape. After some time, however, the cats inevitably hit the latch and were permitted to feed after leaving the box. During repeated exposures the cats became more efficient, requiring progressively shorter time to hit the latch on subsequent trials. Clearly, the cats had learned a response (hitting the latch), as evidenced by the fact that they executed it immediately upon being replaced in the box. This kind of learning is called instrumental conditioning because the animal's behavior is instrumental in obtaining the desired outcome (food).

Aversive Conditioning

Other important examples of instrumental conditioning are those employing aversive stimuli. An example of aversive instrumental conditioning might involve the administration of shock, to the subject through the grid floor of a cage, which the animal may terminate by pressing a small lever protruding into the cage. At first, the animal jumps about vigorously trying to escape the shock and, as in Thorndike's experiment, inevitably hits the lever (which terminates the shock). After a short time, the process is repeated and it is observed that over training, the animal learns to terminate the shock immediately upon its presentation. The learned response is instrumental in escaping from the painful shock.

Operations and Reinforcement

The basic operation which defines instrumental conditioning is that the subject's behavior determines the outcome (presentation of food as in Thorndike's study, or shock termination as in the aversive conditioning example). The rewarding outcome is contingent upon the response being made by the subject; thus the events in the experiment are not independent of the subject's behavior as they are in Pavlovian conditioning.

A response that produces a rewarding outcome will be learned, i.e., it becomes more frequent or probable with continued training. This process is the acquisition phase of learning. If the response ceases to produce the reward, the behavior consequently diminishes in vigor and decreases in probability. This process is called extinction. Thus, these two terms mean the same here as they do in classical conditioning.

Thorndike considered instrumental learning a gradual, "stamping in" process. The initial random movements were eliminated (i.e., were extinguished) because they did not lead to the food reward, while the correct response of unlatching the door was gradually strengthened over training. Thorndike's Law of Effect stated that when a response is reinforced (followed by a satisfying state of affairs which the animal does nothing to avoid but, rather, tries to obtain), it will become stronger in the sense that it will tend to be repeated in the future. The basic notion behind the Law of Effect (responses that are rewarded will be repeated) is the cornerstone of instrumental conditioning.

In more general terms, a reinforcer may be defined as a stimulus that increases response probability. On the one hand, positive reinforcement occurs when the increase in response probability is achieved by the presentation of a positive stimulus (e.g., food as in Thorndike's experiment). On the other hand, negative reinforcement occurs when the response is strengthened by the termination of a negative or aversive stimulus (e.g., shock offset as in the aversive conditioning example). In both cases, reinforcement leads to an increase in response probability.

The change in the probability of a response, then, indicates its relative potency. Frequency, latency, or speed of response are also used to measure response strength. The observation that a rat in a maze takes increasingly less time (or makes fewer errors) to reach the goal box with continued training leads to the inference that the strength of the response is increasing. A more typical measure in instrumental conditioning is the rate of responding (e.g., number of lever presses per minute), which provides an accurate estimate of the response strength or probability.

Instrumental conditioning may be schematically diagrammed as in Figure 3-1. There exists an environment consisting of many stimuli

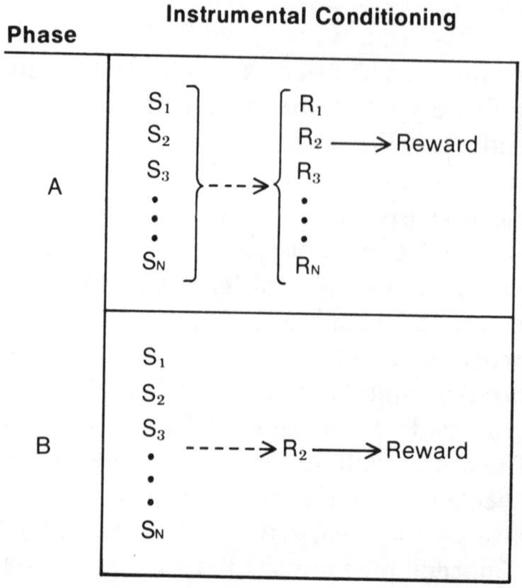

Figure 3–1. Acquisition phases in instrumental conditioning.

(labeled $S_1 \ldots S_n$). In phase A, no one stimulus elicits an observable, reflexive-type response as the US does in classical conditioning. Rather, many different responses (labeled $R_1 \ldots R_n$) are given (e.g., the random, clawing responses in Thorndike's experiment) until one of them is followed by the reward. After sufficient repetition of this sequence, a conditioned response is established (phase B) so that presentation of the same stimulus complex will evoke the response.

Terms

Unconditioned (Reinforcing) Stimulus

As in classical conditioning, an unconditioned stimulus is used as the reinforcer. In this case, however, it is given only *after* the desired response has been executed. In Thorndike's study, the US was food while in the aversive learning example it was shock offset.

As stated previously, a reinforcer may be any event that increases the response probability. Food, of course, is commonly used (as is shock termination) because it is easily administered and controlled. However, many other biological stimuli have been used (e.g., water, sugar-flavored substances, heat for a cold animal, access to a sexual partner). Other nonbiological reinforcers have also been used (see Chapter 10 for a more complete discussion of the nature of reinforcers) including access to an exercise wheel, the opportunity to visually inspect the territory outside a cage, and sweet-tasting but nonnutritive substances.

In summary, then, reinforcement in instrumental conditioning is the presentation of a US (or the removal of a noxious US) following the correct response. While the US in instrumental conditioning is usually a potent, biological stimulus like food, other stimuli will also reinforce or increase response probability.

Conditioned (Discriminative) Stimulus

While in classical conditioning invariably there is an explicit CS, this is not necessarily true of instrumental conditioning. As implied in Figure 3-1, the stimulus situation or complex doesn't need a salient cue (light or tone) present. All of the existing environment may function as a CS (e.g., no lights or tones were presented to Thorndike's cats).

However, an explicit cue may be given in instrumental conditioning. Such a cue is called a discriminative stimulus (S_d). Because the execution of the response is determined voluntarily by the subject, the discriminative cue may only indicate when reinforcement is or is not available. The discriminative stimulus is not necessarily followed immediately by the US (as in classical conditioning) but rather the US follows the S_d *only* if the subject responds. Therefore, the discriminative cue only informs the subject about the availability of reinforcement.

A discriminative stimulus signaling the availability of reinforcement is labeled an S_d, while one signaling a condition where no reinforcement is available is termed an S_Δ.

Unconditioned Response

The unconditioned response is often anomalous in instrumental conditioning to the extent that it bears no resemblance to the CR. In other words, although each US does elicit a UR as in classical conditioning (food elicits salivation, shock elicits motor and visceral responses) it does not always figure prominently in the instrumental conditioning experiment. The unconditioned response to food is certainly not lever-pressing, and while vigorous running or flight may be a UR to shock, other responses are often designated as the instrumental response, e.g., lever-pressing. In summary then, US's do elicit UR's in instrumental conditioning but the UR's are not usually related to the instrumental response being tested.

Conditioned Response

The CR in instrumental conditioning is the response evoked by the S_d (or the stimulus complex in general if no explicit S_d is identified) and followed by the reinforcer. It usually is a voluntary, motor response like lever-pressing. However, the CR ostensibly may be any response which the experimenter designates as correct including such visceral responses as heart rate or blood pressure. Furthermore, as stated above, the CR seldom resembles the UR. This arbitrary, acquired response may be strengthened either through positive or negative reinforcement.

Basic Conditioning Paradigms

Because a detailed account of the various instrumental procedures is given later, only a cursory introduction of these paradigms is presented here. Basically, there are four major types of instrumental conditioning experiments, as represented in Figure 3-2. In two of the cases, the reinforcer given for the response is a positive one, an

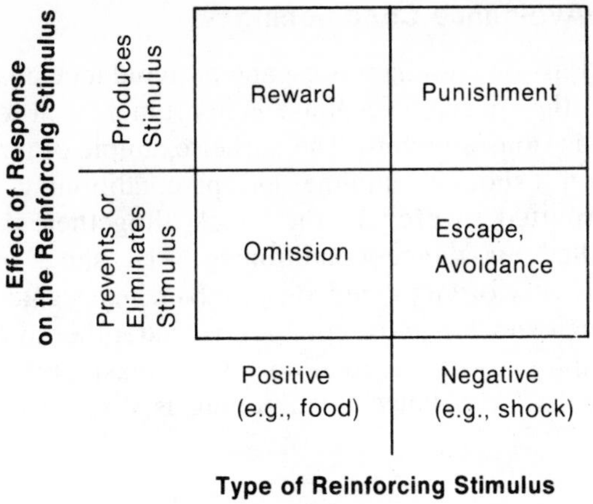

		Positive (e.g., food)	Negative (e.g., shock)
Effect of Response on the Reinforcing Stimulus	Produces Stimulus	Reward	Punishment
	Prevents or Eliminates Stimulus	Omission	Escape, Avoidance

Type of Reinforcing Stimulus

Figure 3–2. Four basic types of instrumental conditioning.

appetitive stimulus like food or water. The two other cases of conditioning use a negative or aversive US, e.g., shock. One type in each of those two conditions involves the presentation of the US while the other type involves the termination (or prevention) of the US.

Reward Training

Reward training is the most common type of instrumental learning. Any trial-and-error procedure in which the subject receives positive reinforcement for making the correct response is an example of reward training. Cats escaping a puzzle box to obtain food, rats pressing levers to get fed, or dogs "shaking hands" to get a biscuit are all examples of reward conditioning.

An important concept in reward conditioning is shaping, a technique for hastening the conditioning session. The purpose of shaping is to train the desired response gradually by requiring closer and closer approximations to the actual correct response until the correct response itself is finally made. For example, in training a rat to press a lever for food, first the rat is reinforced for simply being on the side of the cage where the lever is located. Once that is established, the rat is rewarded for coming close to the lever, for sniffing it, or perhaps biting or pawing it. Once that behavior appears, the experimenter would wait until the bar was pressed before reinforcing the animal. Shaping is not a necessary or essential feature of an instrumental learning experiment since an animal will ultimately learn the correct response if not shaped (as did Thorndike's cats) although the conditioning may take longer to achieve.

Escape-Avoidance Conditioning

A second major paradigm is escape or avoidance conditioning. In this situation, the specified response is instrumental in terminating or preventing a noxious stimulus. The earlier example of a rat pressing a lever to turn off a shock exemplifies escape conditioning. If the animal had been permitted to prevent the shock altogether, the procedure would be called avoidance conditioning. The shock in avoidance conditioning is very often preceded by an S_d which signals the pending shock. Although shock is most commonly used, other aversive stimuli have been employed such as loud noises, air blasts, bright lights, and cold water baths. Avoidance conditioning is discussed at length in Chapter 5.

Punishment

Punishment, a third conditioning situation, occurs when the subject's response produces an aversive stimulus. It is analogous to reward training except that the outcome is negative. In general, the effect of punishment is suppression of the response on which the aversive stimulus was contingent, although there are exceptions to this rule. As in reward training, specific discriminative stimuli may be given (the S_Δ signals the period when the punishment contingency is in effect). See Chapter 6 for a detailed discussion of punishment.

Omission Training

Omission training has received much less attention than the other paradigms. It is defined as a situation in which the response prevents the presentation of an appetitive or positive US, i.e., positive reinforcement is contingent upon *not* responding. The general effect of omission training is suppression of the criterion response, or, in other words, an increase in nonresponding.

There are numerous examples of omission training for voluntary motor responses (see Coughlin, 1972, and Leitenberg, 1965 for reviews) although the term omission training is rarely used. Rather, the time period during which reward is not available is more often called the time-out period. In this situation, a response given in the presence of the S_Δ is not only unreinforced, but also initiates a time-out period during which the apparatus lights are turned off, requiring the animal to wait until the next S_d period. The effect of this time-out period is suppression of responding during the S_Δ. For example, Thomas (1968) trained pigeons to peck a small plastic disc in order to receive food. Initially, after the fiftieth response, a 30-second time-out period followed; the lights in the experimental chamber were turned off and no reward was given for responding. After several sessions, the number of pecks that produced a time-out was reduced to 25, then to 10, and finally to 2 (every other response). Thomas found that the fewer responses required to produce time-out, the greater the suppression of responding, as shown separately for each of three subjects in Figure 3-3. The ordinate scale is a relative one reflecting suppression (.5 indicates no suppression whereas 0 is complete suppression of behavior). It is clear that responding in each subject was only mildly affected when 50 responses produced time-out but was increasingly suppressed as the number was reduced. Furthermore, as indicated on the right side of Figure 3-3, basically no difference was found between a 30-second and 2-minute time-out period, although other studies have found suppression to vary with time-out duration (see Ferster & Appel, 1961). In general, it may be concluded that the effects of time-out from positive reinforcement are similar to punishment; in both cases, responding is suppressed.

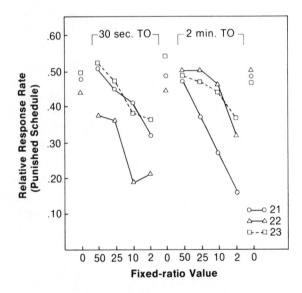

Figure 3–3. Suppression ratio in response to a time-out period for three subjects as a function of the number of responses required to produce the time-out period and the duration of time-out.

It should be pointed out that extinction is essentially an omission procedure, except that it does not usually include S_d periods where reinforcement is available. Extinction, like time-out procedures, suppresses behavior (see Chapter 7 for a discussion of extinction).

Simultaneous Occurrence of Classical and Instrumental Conditioning

An extremely important point is that *both* classical and instrumental conditioning take place in every experiment. Every conditioning situation involves the presentation of contiguous stimuli (classical) as well as changes in the stimuli presentation due to the subject's behavior (instrumental). For example, in instrumental appetitive conditioning, when a rat presses a lever to obtain food, the S_d (or the entire stimulus complex) is at the same time being paired with the food (US) to establish a classical CR. Similarly, in classical defense conditioning, there are motor responses made by the subject which help to minimize the severity of the shock (US). Classical conditioning (CS-US pairing) occurs in every instrumental conditioning experiment, and instrumental conditioning (modification of the US consequences by the subject's behavior) occurs in every classical conditioning experiment. Although both types of conditioning coexist, usually only one response is measured. Nevertheless, it is important to keep in mind that an instrumental S_d may also be a Pavlovian CS and that the US not only conditions a classical response but also modifies the probability of an instrumental response. Many investigators have attempted to isolate the classical or instrumental elements in order to study their role in the conditioning situation. In light of the above point, the success of these studies in creating a pure classical or instrumental conditioning situation is questionable. In contrast, many psychologists have combined classical and instrumental elements in a single experiment in order to study the interaction of the two processes.

Reward Conditioning

Amount of Reinforcement

The magnitude of the reinforcer has been studied in a variety of situations, and it is clear that performance improved with larger rewards. Kintsch (1962) reinforced thirsty rats with different volumes of water (3.5, 1.75, or .25 cc) for traversing an alley. Mean running speed, as shown in Figure 3-4, indicated that performance improved with greater reward. This finding has also been observed when varying quality of reward. For example, Goodrich (1960) and Kraeling (1961) have shown that running speed increases with higher concentrations of sucrose-water solutions.

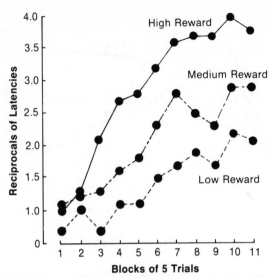

Figure 3–4. Mean start-box speed (reciprocal of latency) for high-, medium-, and low-reward groups as a function of training trials.

Contrast Effects

If, during acquisition, the magnitude of the reward is changed, appropriate shifts in performance are found. Performance improves if the reward is increased; it deteriorates or decreases if there is a decrease in reward. In addition, Crespi (1942) also noticed that the groups for which the amount of reward was shifted to a new value tended to overshoot the performance of a group that had received that amount throughout. This overshooting is called a contrast effect. The general change in performance and the effect of overshooting occurs either when reinforcement magnitude is changed while the animal continues to respond in the same situation (successive contrast), or if the subject is given large and small rewards mixed together during training but in separate apparatus (simultaneous contrast).

In an example of the latter situation, Bower (1961) trained three groups of rats to traverse an alley for food reward for 128 trials. One group received eight food pellets throughout training, while a second control group received one pellet throughout (but in a different alley). The experimental group received the large reward on half of the trials and the small reward for the other half (in the separate alley). Figure 3-5 shows that the high-reward group ran faster in the alley than did the low-reward group, thus confirming the generalization stated previously. The experimental group ran as fast as the controls in the high-reward alley, but much slower in the low-reward alley. In fact, their

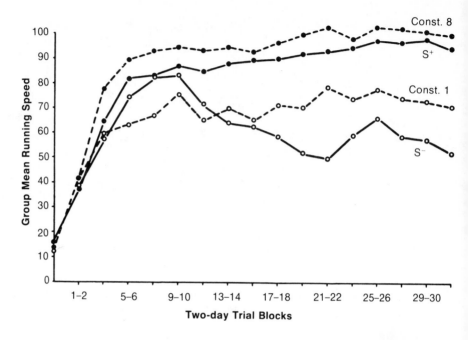

Figure 3–5. Mean running speed as a function of training trials for a large-reward group (const. 8), small-reward group (const. 1), and an experimental group that received both large (S^+) and small (S^-) reward.

behavior in the latter case was depressed below that of the low-reward control group. Thus, Bower obtained a negative contrast effect but not a positive contrast effect. This asymmetry (negative but not positive contrast) has been found frequently (DiLollo, 1964; Spear & Spitzner, 1966) and has been interpreted by Black (1968) in terms of the summation of excitatory and inhibitory effects of reward change.

Both positive and negative contrast effects have been more frequently found when the frequency of reinforcement is changed in a lever-box situation (Mackintosh, Little, & Lord, 1972) as opposed to a change in the amount of reward for an alley response. However, Padilla (1971) found symmetrical contrast effects for both changes in amount and frequency, suggesting that experiments like Bower's (1961) had encountered a ceiling effect: The responding was compressed at the high end of the measurement scale, thus preventing positive contrast effects from being accurately detected. In summary, Padilla claimed that both positive and negative contrast effects can occur and further proposed that these effects represent subjects' emotional reactions. Like Crespi (1942), Padilla called the positive contrast effect "elation" and the negative contrast effect "depression."

Incentive

The contrast or shift studies reveal an important concept in learning—incentive—which was stressed by Spence (1956). Incentive refers to the motivational aspects of the goal object. Spence claimed that a reward exerted an attractive or enticing force on a subject. Whereas a biological need may push or goad the subject into action, the goal object attracts or pulls the animal. Thus, rewards increase responding both by reinforcing as well as by motivating the subject.

The shift experiments clearly demonstrate the effect of incentive by showing that performance improves when the reward is increased in size. Why should this improvement take place if the animals have already learned the response and are performing at their peak? The answer is that with a larger reward, they are not learning an additional amount, but rather they are more enticed or motivated to perform.

Similarly, the latent learning studies, mentioned in Chapter 1, illustrate the incentive concept. Animals learned the maze in Blodgett's (1929) experiment although they did not perform in a manner which reflected their learning. On later trials, however, once a goal object (food) was provided, the subjects immediately performed the correct response. The food didn't suddenly teach the animals the response; rather, it simply motivated them to perform.

The motivational properties of goal objects are now recognized as important determinants of performance. In fact, some theorists (e.g., Walker, 1969) claim that the concept of incentive better accounts for performance changes than does the concept of reinforcement. That an incentive entices the subject to respond is a more logical and useful explanation of performance than that provided by the concept of reinforcement.

The important point is that behavior is modified by means of a change in motivational level. The properties of a reward object, such as its magnitude or immediacy, define the parameters of motivation. The principles pertaining to incentive motivation, then, are the same as those pertaining to learning.

How do reward objects affect motivation? Spence proposed a theoretical mechanism to explain the motivational effects of a goal object on behavior. Stimuli in the goal box first become conditioned to the response of eating. Stimuli in the start box, because they are somewhat similar to the goal-box cues, are able to elicit fractional goal responses (e.g., licking, chewing responses) which add to the vigor of the total response. These fractional responses (r_g) produce their own proprioceptive or internal stimuli (s_g) which become part of the overall stimulus complex eliciting the running response.

The proposed operation of the incentive mechanism is shown in

Figure 3-6. Goal stimuli are conditioned to the goal response of eating. Subsequently, similar stimuli in the start box not only elicit the instrumental response but also the r_g—s_g sequence. As a consequence, *both* the stimuli at the start and the s_g's elicit the response of running.

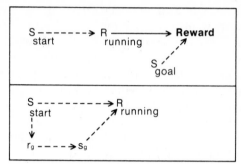

Figure 3–6. Diagram of the incentive r_g—S_g mechanism. The start-box stimuli are similar to the goal stimuli and thus elicit fractional, anticipatory goal reactions. The resulting response-produced stimuli form part of the complex that evokes the running response.

Delay of Reinforcement

A second important parameter in instrumental conditioning is the delay of reward. This variable is similar to the CS-US interval in classical conditioning. In general, the effect of increasing reward delay is a decrease in performance. Furthermore, a systematic gradient is found, asymptotic performance being inversely proportional to delay.

Early studies showing this basic effect attempted to measure the shape of the gradient, the extent to which the reward could be delayed and yet still allow the subject to learn the response. Wolfe (1934) postulated that a delay of 2 minutes was the limit to which a subject could be exposed and still learn the correct turn in a T-maze; however, Perin (1943) found that 30 seconds was the maximum delay which still allowed for the acquisition of a lever-press response. It was assumed that the stimulus-response "trace" persisted for some time and was strengthened by the reward if it was presented soon enough (before the trace had dissipated completely). Thirty seconds was thought to be the limit over which the trace would persist, although Wolfe's study was at variance with that notion.

Hull (1943) resolved this problem by claiming that Perin's gradient showed the decay of the trace and, therefore, the limits over which reward was effective, while Wolfe's gradient included the effect of an additional factor. This factor was the action of the apparatus cues (e.g., smells, goal-box color, sight of the food cup) which existed in the alley and goal box and helped the subject to bridge the delay and thus sustain its behavior. These cues were called secondary reinforcers (for a full discussion of secondary reinforcement see Chapter 8). In summary then, Hull believed that the gradient of delay extended to 30 seconds unless additional reward-related cues were present which would sustain the behavior in the temporary absence of food reinforcement.

Hull's view was challenged by Grice (1948), who used a special task that minimized the effect of the secondary cues mentioned above. Grice found that the limit for reinforcement delay was 5 seconds; virtually no learning occurred with longer delays. Such experiments led Spence (1947) to conclude that *all* learning which involves delay of reward stems from the action of these secondary cues. Spence's hypothesis has received considerable support, principally from studies which showed that the addition (or subtraction) of these secondary cues increased or decreased delayed-reward learning (e.g., Grice, 1948; Renner, 1963; Tombaugh & Tombaugh, 1971). In conclusion, the gradient appears to be very steep; performance deteriorates quite significantly with even small delays, but the presence of secondary cues does allow the subject to learn under much longer delays (for reviews of the delay of reinforcement literature, see Renner, 1964, and Tarpy & Sawabini, 1974).

Other interesting effects occur when the reward is delayed. Pubols (1962) ran subjects in a T-maze: Responses to one side were followed by a constant delay of reward (5 seconds for one group, 15 seconds for another) while responses to the other side were followed by a variable delay of reward (no delay or twice the constant delay). Thus, the average delay was the same for both responses, but the variability of delay differed. Pubols found that the rats preferred the variable side and, in a second study, the preference developed more quickly with longer average delays. The results imply that with variable delay, the beneficial effect of immediate reinforcement on behavior counteracts and outweighs the deleterious effect of the long delay.

The relative effects of delay have also been studied (Logan, 1965). Hungry rats were given a choice between traversing one of two alleys, one of which led to a larger magnitude of reward in the goal box than did the other, but the delivery of the reward was delayed. Improvement in performance due to greater magnitude would, in turn, be reduced by the added reinforcement delay. Logan systematically varied both the amount and delay and thereby was able to accurately measure the effects of delay, i.e., how many seconds' delay was required to offset a given amount of food. He derived functions depicting the values of delay and amount which were equivalent in their effect (but in the opposite direction). This study is important in illustrating a quantitative approach to the study of delay as it interacts with other variables.

Finally, contrast effects have been observed after a shift in delay. For example, McHose and Tauber (1972) gave 30 trials in a straight alley to two groups of rats. One group was given 5 reward pellets in the goal box after 10 seconds delay; the other group received a 30-second delay. Next, all animals were given 48 more trials during which half of each group was shifted to the other delay value and half continued to receive the same delay. As shown in Figure 3-7, an increase in delay (group 10-30) produced negative contrast, a depression effect, whereas a reduction in delay (group 30-10) did not produce the corresponding positive contrast effect (elation). These results parallel much of the work on shifts in the amount of reinforcement insofar as a negative but not positive contrast effect was found (Black, 1968). McHose and Tauber (1972) concluded that delay and amount of reinforcement may be fundamentally related to the extent that they effect behavior oppositely but via a single mechanism—incentive motivation.

Drive Level

Drive is a hypothetical variable reflecting the motivational state of the organism. Usually, drive is specified by the operations which produce it, such as, hours of food deprivation, percent of body weight

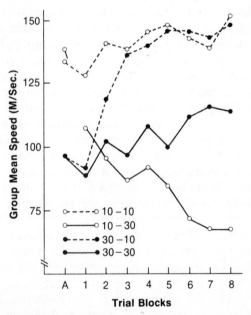

Figure 3–7. Mean running speed as a function of trial blocks following a shift (groups 10–30, 30–10) or no shift (groups 10–10, 30–30) in delay of reinforcement.

loss, or, in the case of aversive conditioning, shock intensity. Numerous studies have investigated the effect of drive level on learning and/or performance. The general finding is that asymptotic performance is higher with greater drive (e.g., Kintsch, 1962; Pavlik & Reynolds, 1963; Zaretsky, 1965). In Zaretsky's experiment, different groups of rats were deprived for either 22 or 1.5 hours prior to being trained in a straight alley. The effect of increased deprivation on acquisition was much higher asymptotic performance, as shown in Figure 3-8. This indicates that with greater motivation, animals perform at a higher level.

It is possible that animals perform at a higher level not only because drive affects performance, but also because high drive leads to better learning (greater response strength). This question has been investigated by shifting drive states and observing whether performance is better following high drive than following low drive. For example, two groups of animals are trained to make a response, one under high-drive and the other under low-drive conditions. Both groups are then shifted to the same medium-drive level. If drive only affects performance, then both groups should perform equally well. However, if learning is affected by the previous drive level, then the former high-drive group should perform better than the former low-drive group. Several experiments have shown that the latter is the case. Initial motivational levels do influence subsequent performance even if the drive conditions have changed (e.g., Butter & Campbell, 1960; Capaldi, 1971; Zaretsky, 1966). These findings, then, suggest that drive affects learning as well as performance.

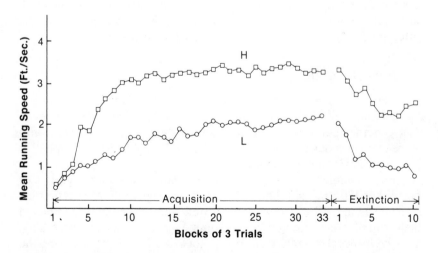

Figure 3–8. Mean running speed in an alley during acquisition and extinction for groups deprived of food for either 22 hours (H) or 1.5 hours (L).

When testing the effect of a previous drive level on extinction performance, however, the results are less clear-cut. Although Campbell and Kraeling (1954) and Theios (1963) found that extinction was prolonged by higher previous hunger levels, Pavlik and Reynolds (1963) and Leach (1971) have not. The reason for such a discrepancy is not entirely clear. Although changes in drive do not seem to interact with changes in reward magnitude (Capaldi, 1971; Pavlik & Reynolds, 1963), the influence of a previous drive on current behavior may depend on the extent of training. For example, Mollenauer (1971) shifted deprivation levels after 23 trials in an alley and found gradual changes in performance, indicating a persisting effect of the previous drive state. However, when the shift occurred after 75 or 105 trials, no residue effect of the prior drive state was observed. Instead, the rats immediately shifted their performance and, in fact, demonstrated both positive and negative contrast effects. As is true with other variables, the effect of drive on learning and performance is surely not unitary. Rather, drive is one of a number of variables which interact in a complex fashion to influence behavior, and additional research is necessary before the nature of these influences is clearly understood.

Escape Conditioning

A considerable amount of research has also been done on instrumental escape conditioning, in which the subject performs an instrumental response to turn off a noxious US. The most popular stimulus has been electric shock because it is easily specified and administered, although, more recently, other aversive stimuli are gaining wider attention, e.g., loud noise, heat, cold, cold water, bright light.

Intensity of US

Perhaps the most important parameter of escape conditioning is the US intensity, primarily because it defines the drive (aversion) level. The general finding is that acquisition and extinction are faster with more intense shock (e.g., Bower, Fowler, & Trapold, 1959; Campbell & Kraeling, 1953; Staveley, 1966).

For example, Trapold and Fowler (1960) gave separate groups of rats either 120, 160, 240, 320, or 400 volts (through a 250-K ohm limiting resistor) as the US which was terminated when the animals completed their run down a straight alley. Running speed was an increasing, negatively accelerated function of shock intensity; the stronger the shock the faster the subjects ran, as shown in Figure 3-9. The implication of this study is that escape performance is related to drive

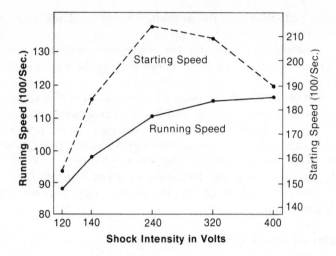

Figure 3–9. Mean running speed (left ordinate) and mean start-box speed (right ordinate) as a function of shock intensity.

(shock) level in the same manner that reward training is related to, say, the level of food deprivation.

More recently, Franchina (1969) has shown that subjects who receive varied shock intensity perform better than comparable groups that receive only one intensity. In this study, rats could jump over a hurdle to terminate shock. Three different groups received 20, 50, or 80 volts respectively (constant dc voltage source) while a fourth received all voltages in random order. Speed was positively related to shock intensity; furthermore, the group that received all intensities performed better than the other groups.

Amount of Reinforcement

Traditionally, the amount of shock reduction in escape learning has been analogous to reward magnitude in reward training. Campbell and Kraeling (1953) were the first to investigate this parameter by varying the percentage of reduction in shock following the escape response. Rats were required to traverse a straight alley to escape a 400-volt shock (250-K ohm limiting resistor). Shock in the goal box was reduced by either 25, 50, 75, or 100 percent, i.e., a 100, 200, 300, or 400 volt reduction. The results showed that the percentage reduction was important, with the greater percentage decrease leading to better performance. Thus, the greater proportional reduction in drive, the greater reinforcement (and performance). Using other initial levels of shock, Campbell and Kraeling were able to show that it was the percentage reduction, not the absolute change in shock, that defined the reinforcing dimension. A given percentage reduction, say, 50 percent, was always equally reinforcing regardless of the initial shock level. These results have been confirmed by Bower, Fowler, and Trapold (1959).

A second aspect of the issue of reinforcement magnitude is the duration of time produced by the escape response or the intertrial interval. That a longer safe period following the response would produce better performance was shown by Staveley (1966). When rats could press a lever to terminate shock for either 0, .5, 2, 8, 32, or 130 seconds, speed of response increased as a function of the intertrial interval. Staveley also found that the improvement was magnified at higher shock intensities.

Delay of Shock Offset

As in appetitive conditioning, delay of reward (delay of shock offset) has a pronounced effect on escape performance: With increasing delays, response speed is much slower. Fowler and Trapold (1962) allowed rats to traverse a straight alley to terminate shock, the offset of which was delayed either 0, 1, 2, 4, 8, or 16 seconds for different groups. Running speed (Figure 3-10) was inversely related to delay.

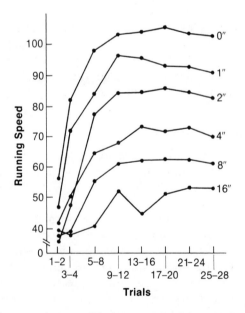

Figure 3-10. Mean running speed during escape training as a function of the delay of shock offset.

Fowler and Trapold found that the gradient extended to at least 16 seconds (the 16-second group did learn the task). It is clear, however, that the gradient is severely restricted when the available secondary cues are reduced. (This was noted previously to be true for appetitive conditioning as well.) In fact, rats are extraordinarily sensitive to delay of shock offset and, unless cues are provided, even small delays preclude learning.

That rats are very sensitive to delay of shock offset was shown in a study by Tarpy (1969). Rats could press either of two levers to terminate shock. Shock offset was delayed for both responses, but the delay values differed. Rats developed a preference for pressing the lever that produced the shorter delay; furthermore, the differences in delays required to produce this preference were extremely small. For example, a preference was shown for the 1-second delay lever versus the 1.3-second delay lever. This finding indicated that performance is adversely affected by far smaller delay values than previously noted.

The second point, that even small delays (e.g., 3 seconds) preclude learning unless secondary cues are provided, was demonstrated by Tarpy and Koster (1970). In this study, rats pressed a lever to terminate shock. For one set of groups, shock offset was delayed either 1, 3, or 6 seconds, and no stimuli were given during the delay interval. For the other three delay groups, a small light was turned on during the delay interval. As shown in Figure 3-11, the no-cue groups were inferior to

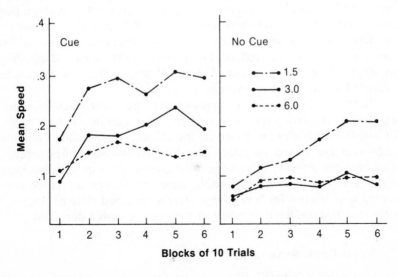

Figure 3–11. Mean speed during escape training as a function of delay of shock offset and cue presentation during the delay period.

the corresponding cue groups. In fact, without the light cue, learning with even a 3-second delay of shock offset was negligible. Thus, the Tarpy and Koster experiment shows a much steeper gradient than the one found by Fowler and Trapold largely because there are fewer cues in a lever box that are differentially (exclusively) related to shock offset, as compared to the goal box of an alley where such cues do exist. The more general conclusion is that escape learning resembles appetitive conditioning in that delay of reinforcement profoundly retards performance and that the addition of cues during the delay interval facilitates the behavior.

Schedules of Reinforcement

All acquisition situations are characterized by a pattern and relative frequency or schedule of reward (to this point, only continuous reinforcement has been discussed). Because a subject's performance is profoundly affected by reinforcement schedules, the most widely investigated parameter of reward conditioning has been the reward schedule (see Schoenfeld, 1970, for a review).

Rate of response is the typical measure when investigating reinforcement schedules. However, because subtle changes in the rate cannot be observed if the responding is averaged over, say, an hour, techniques have been devised to record moment-by-moment rate of responding. These techniques produce a plot of the responses per unit time called a cumulative record. Each time a response occurs, a pen moves one step toward the top of a paper which is moving at a constant rate from right to left. Therefore, if no responding occurs, a flat horizontal line is obtained on the paper; if very rapid responding occurs, a steep line is obtained. Rate of responding, then, may be measured at any given moment in training.

There are four basic schedules of reinforcement, each of which specifies an operation for delivering reward on an intermittent basis. Two of the schedules are based on the subject's responding while the other two are based on time. The former two schedules arrange reinforcement delivery after a specified number of responses has been made (either a fixed or a variable number). Under the latter two, reward is delivered for a response after a specified time has elapsed (either a constant, fixed duration of time or a variable duration).

Fixed-Ratio Schedules

Under the fixed-ratio (FR) schedule, the subject receives reward for each Nth response. After a fixed number of responses is given, the next response will produce reward. The simplest FR schedule is the alternating pattern where reward is presented for every other response. However, FR schedules may involve a much larger number of responses.

In general, the performance under a FR schedule is very high and stable, although species differ according to their rate at each fixed-ratio value (FR-10 is about optimal for a rat, while FR-50 is optimal for a pigeon). A typical cumulative record illustrating the pecking behavior of a pigeon on a FR-50 schedule is shown on the left side of figure 3-12.

The reason for rapid responding is that the subjects themselves determine the frequency of reward (the faster they respond, the sooner the reward is delivered) and therefore are reinforced for fast responding as opposed to slow responding. The stability of responding is related to the size of the ratio. When the ratio of responses to reinforcements is small, the rate is very stable; however, with large ratios, a significant postreinforcement pause is evident. For example, Felton and Lyon (1966) found no postreinforcement pause for pigeons under a FR-50 schedule (note the straight lines in Figure 3-12) while large pauses (indicated by arrows) were obtained with a FR-150 schedule (right-hand side of Figure 3-12).

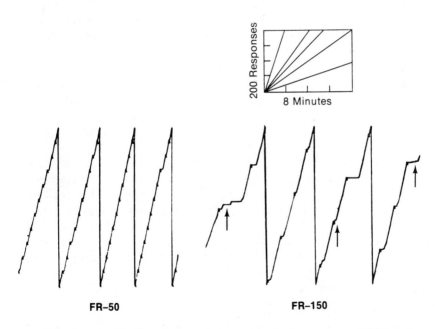

Figure 3-12. Cumulative record for a pigeon responding under a FR-50 and a FR-150 schedule of reinforcement.

Variable-Ratio Schedules

The variable-ratio (VR) schedule is similar to the FR except that the number of responses required before reinforcement is delivered varies randomly. Responding under this schedule is normally very high and stable with very few pauses even with large response-reinforcement ratios (although there is an upper limit to the ratio size before pauses start to appear). As in the case of FR schedules, the rate is high because fast responding produces quicker reinforcement than slow responding. Stability is achieved, in part, by the variability of the schedule which prevents any obvious patterns from becoming established.

Fixed-Interval Schedules

In a fixed-interval (FI) schedule, reinforcement is delivered to the subject for the first response after a fixed amount of time since the last reinforcement has elapsed. This consistent pattern allows the animal to anticipate when reinforcement is pending, and the result is an increase in rate toward the end of the interval. This scalloping effect is seen in Figure 3-13, which illustrates the pecking behavior of pigeons on a FI-2 minute schedule.

Immediately after reward, the animal virtually stops responding but the rate increases steadily toward the end of the interval. One

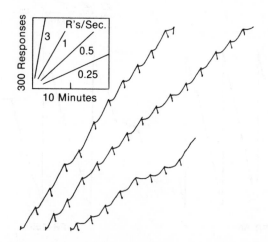

Figure 3-13. Cumulative record for a pigeon responding under a FI-2 minute schedule of reinforcement.

major difference between the FR and FI schedules is that, in the former, rate of responding is constant although there may be pauses. In the FI schedule, however, the rate varies throughout the interval because fast responding is not reinforced (as it is in the FR schedule) since the immediacy of reward is not dependent upon the subject's rate of responding.

According to Dews (1962), the responses made during the interval do not represent a chain of responses with one response leading to the next. Rather, the performance at any given time depends on the temporal relationship between the response being made and the subsequent reinforcement. In view of the findings on delay of reward, responses executed early in the interval would not be as strong as those given late because reinforcement for the early responses would be relatively more delayed. Therefore, delay of reinforcement is responsible for producing the scallop effect in which rate of responding during the interval is inversely related to reward immediacy.

Variable-Interval Schedules

The fourth basic schedule is the variable-interval (VI) schedule. Reward is available after a specified time has elapsed, but the duration is randomly varied which precludes any consistent pattern for anticipating pending reinforcement.

The characteristic rate of responding under a VI schedule is stable but much lower than under a VR schedule that yields a comparable reinforcement-per-minute value. Again, the reason for this is the fact that reinforcement frequency (immediacy) is not related to the subject's responding rate. In fact, responses that occur after a relatively long interval are more likely to be reinforced (since longer intervals between responses bring the subject closer to the time of pending reinforcement) than responses that follow quickly. Thus, under a VI schedule, although the responding is steady, the interresponse times are much longer, and as a consequence, the overall rate is lower.

DRL Schedules

An interesting situation, although not a basic schedule, is one in which the subject is explicitly not reinforced for responding quickly. Rather, reward is given only if a certain time has elapsed since the last response. This schedule, similar to omission training discussed earlier, is called differential reinforcement of low rates of responding (DRL). The subject must withhold the response for, say, 30 seconds, before reward becomes available. Initially, the behavior fluctuates; a reinforcement increases response rate which leads to nonreinforcement, which in turn decreases rate, and so on until some stability is achieved. Except for some bursts of several responses with very short inter-response times, the average interresponse interval increases with training to the point where the response is finally withheld for the specified time duration. Although Skinner (1938) described a schedule similar to a DRL, it is only recently that attention has been given to this paradigm (see Kramer & Rilling, 1970, for a review).

Complex Schedules

Quite obviously, schedules may be combined in a variety of ways. When reward is contingent not just on elapsed time or number of responses but on the requirements of two distinct schedules at the same time, responding is under the control of a compound schedule. A great number of compound schedules may be established. For example, a compound FI-FR schedule stipulates that reinforcement is available only if the subject executes a specified minimum number of responses within a specified time. If either condition is not met, reinforcement is not delivered. Behavior under such a compound schedule is a combination of the characteristic responding under each individual schedule. Although scalloping appears, the subject responds at a higher rate prior to the end of the interval.

There are several ways of presenting schedules sequentially. The first (called tandem schedule) is when the subject must complete the requirements of each individual schedule in succession before reinforcement will be given. If a separate, external cue signals each individual component schedule, the arrangement is called a chained schedule. In a third arrangement, called a mixed schedule, the subject may obtain reinforcement during each individual component schedule but different schedules are presented sequentially, in a random order. Finally, these component schedules may be signaled with a discriminative cue, in which case the arrangement is called a multiple schedule. Rate of responding is determined not only by the component schedule currently in effect but also by the other component schedules which make up the multiple schedule.

Concurrent schedules involve the reinforcement of two separate responses under two distinct and independent schedules. For example, a pigeon may peck a red disc under a VI schedule or a blue disc under a different VI schedule. In general, the behavior characteristic of each simple component schedule is maintained. Furthermore the subject's response rate tends to reflect or match the reinforcement rate on each component schedule so that rate of responding on a VI-1 minute schedule would be approximately twice that on a VI-2 minute schedule.

Complex schedules are widely used because potentially they provide a more realistic setting for the study of reward schedules and behavior. Very few behaviors are reinforced according to a simple schedule; rather, the contingencies represent the multiple action of several schedules. The analysis of reinforcement schedules has shown that extensive similarities exist between complex human behavior and simpler behavior (e.g., that of a pigeon) despite the obvious differences between the organisms.

Special Cases in Reward Conditioning

Throughout this discussion on instrumental conditioning, an important principle has been articulated: A response on which subsequent reinforcement is contingent is strengthened by the reward. Without the response-reinforcement contingency, no change in performance is observed. The principle of reinforcement contingency, like the principle of contiguity in classical conditioning, seems to be at the core of instrumental learning. This point is graphically illustrated in three special cases in conditioning (although other cases might easily be included).

Superstitious Behavior

Skinner (1948) made a remarkable discovery when he showed the conditioning of a superstitionlike response. Hungry pigeons were put into the apparatus and fed after a fixed time had elapsed. The food presentation had nothing to do with their behavior, yet conditioning took place for six out of eight subjects. One subject learned to turn counterclockwise about the cage; a second thrust its head into an upper corner of the cage; while a third developed a tossing response with its head. In each case the behavior appeared consistently, just prior to the food delivery. The subjects behaved as if these responses caused the food to appear, although no causal relationship was present (thus the label superstitious behavior). Skinner reasoned that whatever response a subject might have been making at the time when food was presented was strengthened by the reward, and, therefore, repeated in the future (but see Staddon & Simmelhag, 1971, for a review).

This is one of the best examples showing the effect of response-reinforcement contingency. The same procedure led to the conditioning of vastly different responses in other subjects simply because the responses and reward accidentally occurred together in succession. Skinner's experiment on superstition may represent the minimum conditions necessary and sufficient for instrumental reward conditioning, just as the sensory preconditioning experiments may represent the minimum conditions necessary for classical association learning.

Verbal Operant Conditioning

A fascinating area of study is verbal operant conditioning in which response-reinforcement contingency is also clearly demonstrated as the basic ingredient in reward conditioning. However, in the case of verbal operant conditioning, the reward is not a conventional one like food or water, and the conditioned response (a verbal response) appears to be learned without the subject's awareness (see Das, 1969, for a review).

Thorndike (1932) used the words "right" and "wrong" to reinforce responses in humans, although verbal operant conditioning is achieved with even subtler reinforcers. For example, in a study by Greenspoon (1955), subjects were asked to say words (not sentences). For one group, the experimenter said "mm-hmm" following each plural noun; no verbal reinforcer was given to the control group. Greenspoon found that the incidence of plural nouns increased systematically with training and that the subjects were unable to verbalize the relationship between the response and the contingent reinforcer.

Additional research has demonstrated that awareness probably does exist in such conditioning but that the subjects' verbal report poorly reflects their structured behavior; thus, what subjects say they know does not always correspond to the way they behave (Verplanck, 1962). Nevertheless, verbal operant conditioning is important, for it illustrates how reinforcement contingencies modify complex verbal behavior.

Instrumental Autonomic Conditioning

The final case to be considered is instrumental conditioning of visceral responses, a rapidly changing field in psychology that has received considerable attention in recent years. The impetus for this research is twofold. First, conditioning of internal, visceral states (as mentioned in Chapter 2 with regard to interoceptive conditioning) has profound implications for the understanding and treatment of psychosomatic disorders. In fact, techniques only recently established in the laboratory are being used in clinical settings. Second, the fact that visceral responses may be conditioned instrumentally at all is a surprise, and thus an important theoretical contribution to the study of learning. Until recently the involuntary, autonomic nervous system was thought to be modifiable only by classical conditioning techniques, while voluntary, motor responses could only be instrumentally conditioned. However, there is some evidence that instrumental techniques will modify even internal, visceral behavior. This finding gives further support to the generality of the principle of response-reinforcement contingency.

Miller and his associates are the ones most directly responsible for the progress in this area (see Engel, 1972, and Miller, 1969, for reviews). In one study by Miller and Banuazizi (1968), rats were first fitted with a brain electrode. If a minute electric current is passed through such an electrode, a pleasurable effect is experienced by the subject such that the subject will learn to press a lever to administer the current to itself. Thus, electrical brain stimulation was used as the reinforcer. The rats were then fitted with instruments that measured intestinal contractions and heart rate. The subjects in group 1 were reinforced (given brain stimulation) each time their heart rate increased beyond the baseline level; group 2 was reinforced for a decrease in heart rate; group 3 was reinforced for an intestinal contraction; and group 4 was reinforced for intestinal relaxation. The subjects received reward at any time if the appropriate response was made, except during 10 test trials (10 seconds each) where the response level was sampled.

The results are shown in Figure 3-14. In the top graph, stomach contractions increased or decreased appropriately for those groups reinforced for such changes. In contrast, groups 1 and 2 showed no changes in their intestinal contractions. The opposite effect is seen for heart rate in the bottom graph. Here, heart rate was appropriately modified by the reinforcement in groups 1 and 2 but not in groups 3 and 4. In summary then, only the particular visceral response, on which the rewarding brain stimulation was contingent, changed while other nonreinforced responses stayed the same.

The above experiment is an example of reward conditioning; however, visceral responses have been instrumentally conditioned with negative reinforcement or have been suppressed with punishment. An example of the former is a study by DiCara and Miller (1968b) on heart rate. Rats were curarized and fitted with surface electrodes for measuring heart rate. Conditioning consisted of the presentation of a 5-second CS (light). If the subject's heart rate increased (or decreased for another group) to the criterion level during the 5 seconds, the shock (administered to the tail) was avoided and a tone came on until the next trial. However, if the heart rate did not change appropriately within 5 seconds, a shock pulse was delivered every 2 seconds until the change finally occurred. The heart rate was measured every tenth trial for 300 trials. The results indicated that visceral responses may be instrumentally conditioned using negative reinforcement (the effect is not limited to reward conditioning for which brain stimulation is the reinforcer).

The degree of response specificity and the control over these responses by reinforcement contingencies is extensive. Furthermore, numerous responses, in addition to heart rate, have been conditioned in rats, monkeys, and humans, including GSR (Kimmel, 1967), salivation (Miller & Carmona, 1967), vasoconstriction (DiCara & Miller, 1968a), and brain waves (Kamiya, 1969).

As mentioned above, instrumental autonomic conditioning has profound implications for the treatment of psychosomatic disorders. In more theoretical terms, the results appear to demonstrate that *all* measurable responses may be influenced by instrumental reinforcement contingencies: Visceral and glandular response modification is

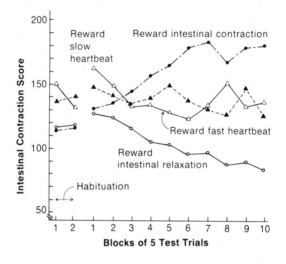

Figure 3–14(a). Intestinal contractions as a function of test trials.

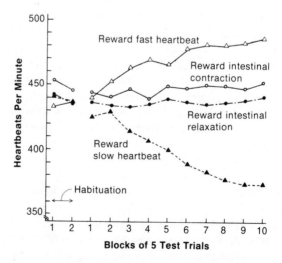

Figure 3–14(b). Heart rate as a function of test trials.

not limited to classical conditioning techniques. One implication is to support the Law of Effect (reinforcement contingency) as the single, most basic principle governing the learning of all responses.

However, are these results true examples of instrumental conditioning? At the present time, the theoretical issue of whether visceral responses may be instrumentally conditioned is far from settled. In fact, some investigators now claim that visceral responses *per se* cannot be instrumentally conditioned, or at least that classical conditioning cannot be excluded as the possible cause for these results.

One reason for such doubt is that some results have not been replicated (a necessary condition before drawing firm conclusions). More importantly, a number of investigators claim that visceral responses cannot be instrumentally conditioned in the absence of skeletal muscle movement (e.g., see Black & de Toledo, 1972). As discussed by Katkin and Murray (1968), the proof that visceral responses can be instrumentally conditioned depends on the elimination of all skeletal muscle movement by curare, a drug which paralyzes the animal's skeletal muscles (necessitating the use of artificial respiration) but otherwise has no effect upon the brain or visceral organs. The use of curare precludes the possibility that the instrumental reinforcer actually strengthens an unnoticed movement (such as a breathing response) which, in turn, reflexively changes the autonomic measure (e.g., heart rate). In other words, if changes in heart rate can be conditioned instrumentally, it must be shown that the reinforcer acts on heart rate, not on a skeletal motor response. Such a skeletal response, of course, can be conditioned instrumentally and would reflexively change the heart rate.

In conclusion, it is not completely clear whether visceral responses can be instrumentally conditioned in the absence of skeletal muscle movement. Therefore, it cannot be said unequivocally that visceral responses *per se* can be conditioned instrumentally (but see Crider, Schwartz, & Shnidman, 1969, and Gavalas-Medici, 1972). Regardless of this issue, though, the experiments in this area have made an important contribution to learning theory by specifying an interesting and potentially useful phenomenon that may ultimately answer many theoretical questions concerning classical and instrumental conditioning.

Summary

To many psychologists, instrumental conditioning represents a distinct learning process. Operationally, instrumental conditioning involves the presentation or withdrawal of a US (either appetitive or noxious) contingent upon the subject's response. Contingent reinforcement leads to an increase in response strength as measured by the probability of the response. It is important to note that all learning situations involve both classical and instrumental conditioning elements: Instrumental reinforcers may also be classical US's; S_d's may also function as CS's.

There are four basic conditioning paradigms: reward conditioning in which an appetitive stimulus is given following a response; instrumental escape-avoidance training in which a noxious US is terminated, or avoided altogether, by the response; punishment in which the response initiates a noxious US; and omission training in which the response prevents the occurrence of an appetitive US. In the first two paradigms, the effect of the reinforcement contingency is to increase the response speed and/or probability, while in the latter cases the contingency decreases response probability.

Of these paradigms, reward conditioning has been studied most extensively. In general, instrumental performance varies as a function of the reward magnitude and may overshoot the level of performance of a control group when reward magnitude is changed, indicating that reward affects performance through incentive motivation. Delay of reward and drive level are also parameters affecting reward training.

Similar variables have been studied with regard to escape conditioning. Performance is better with more intense US's and a greater amount of US reduction while a delay of US termination retards performance. As a rule, many variables affect escape conditioning as they do reward conditioning.

The schedule of reinforcement is a major factor in instrumental training. There are four basic types of reward schedules which may be combined in a variety of ways. On a fixed-ratio schedule, the subject receives reward after a fixed number of responses has been executed. On a variable-ratio schedule, the number of requisite responses varies. Fixed-interval schedules provide reward for the first response following a fixed amount of time. The fourth pattern, variable-interval schedules, is similar except the time lapse, prior to reward availability, varies.

Three special cases illustrate the principle of reward contingency. The first is superstitious behavior where a response was acquired because it was closely followed by reward, even though the response was not necessary to produce the reward. The second case involves the acquisition of verbal responses, such as the use of plural nouns in speech, by the contingent presentation of a reinforcer. The final case refers to the conditioning of autonomic responses by instrumental conditioning techniques. Although numerous examples show that such conditioning is possible, the issue of whether the visceral responses can be instrumentally conditioned is not fully resolved.

Classical & Instrumental Conditioning Compared

Solomon & Rescorla

Introduction

It was pointed out in the first three chapters that classical and instrumental conditioning may represent not only different techniques for training responses, but also different learning processes. Two-factor (or two-process) theorists claim that two distinct learning processes exist, although there is lack of agreement about the laws governing the processes. These theorists verify their claim for two processes by citing several major differences between the classical and instrumental conditioning situations.

Other theorists maintain an opposite position, stating that classical and instrumental experiments do differ on the basis of how they are conducted, but that each represents a different demonstration of a single fundamental learning process.

Actually, there is a third position which suggests that the question of whether one or two learning processes exists need not be asked at all (Skinner, 1950). According to adherents of this position, the reason that the question is fruitless is that semantic problems often cloud the issue and that a final answer cannot be known (since it is performance and not learning that is investigated). Learning, as a process, cannot be validated independently from performance; therefore, it is imperative to study performance without postulating internal, unobservable processes. This point of view will be discussed at greater length in Chapter 10.

For the present, however, the question of one or two processes will be discussed. First, the discussion has heuristic value. More important, many psychologists are interested in speculating about the processes which are manifested in behavior, and have felt that it is useful to differentiate the processes when data and logic allow it. The resultant theory, as stated in Chapter 1, provides a foundation and guide for further inquiry.

Are there two learning processes or is there only one? There is no simple answer to this question. As a useful starting point, the questions concerning what distinguishes classical from instrumental *experiments* (as opposed to processes), and how these experiments interact, are viable ones. Answers to these questions, in turn, may provide a framework for considering underlying processes (see Rescorla & Solomon, 1967, for an excellent discussion of two-process theory).

Distinctions Between Classical and Instrumental Conditioning

Operational Distinction

There are a variety of distinctions drawn between instrumental and classical conditioning which have been used to infer differences in process. The first is the operational distinction. In classical conditioning, the US is presented without regard to the subject's behavior, whereas in instrumental conditioning, the US occurrence is contingent upon the subject's prior response.

Response Distinctions

Historically, one of the most widely cited differences is that classical conditioning is pertinent to conditioning of autonomic nervous system responses, while instrumental techniques operate on skeletal muscle responses. This is a true distinction only on the surface. Heart rate and salivation are usually given as examples of classical conditioning, whereas lever-pressing is typically the example of reward conditioning. However, the generality which limits classical techniques to autonomic responses and instrumental techniques to skeletal responses is not well founded. First, several skeletal responses may be conditioned classically as well as instrumentally, namely, eye blink and limb-flex. Conversely, the recent work (cited in Chapter 3) on instrumental conditioning of visceral responses demonstrates that modification of visceral responses may not be limited to classical conditioning methods. It appears, therefore, that a distinction between classical and instrumental conditioning on the basis of response specificity is not warranted.

This conclusion, however, should not be stated too firmly. Both instrumental and classical responses occur in all learning experiments. Therefore, it is possible that skeletal responses, like the eye blink, are actually being conditioned instrumentally (i.e., they have instrumental consequences such as decreasing the force of the unpleasant air puff) despite the fact that the experimenter believes the change in CR strength is due to a classical conditioning process. Similarly, as discussed in Chapter 3, it is possible that visceral responses, which appear to be conditioned instrumentally, are actually being elicited by skeletal muscle movements and therefore the conditioned autonomic responding is due to a classical, not instrumental, process. In conclusion, even though many issues are not resolved, the differentiation of classical and instrumental processes based upon the skeletal-visceral response distinction is a weak one.

A more general response distinction which has often been used to contrast classical and instrumental conditioning involves the voluntary versus reflexive nature of the response. Classical responses, according to this argument, are involuntary, reflexive, and reliably elicited by a stimulus, whereas instrumental responses are voluntary, not reflexive, and freely emitted. The problem here is that it is often difficult to specify the degree of reflexiveness. Surely lever-pressing doesn't just happen; it is elicited by a particular stimulus (some neural event), but that eliciting stimulus is not obvious to the experimenter as the US is in a classical experiment. What seems more appropriate is Turner and Solomon's (1962) notion that there is a general dimension or scale of reflexiveness on which all responses may be placed according to the ease of eliciting them with a specific, known stimulus. Thus, although salivation would be highly reflexive relative to bar-pressing, there would be no discontinuity on the dimension; classical and instrumental would be equally viable for all responses.

A final response distinction involves the similarity between the CR and the UR. It is claimed that classical conditioning differs from instrumental because the CR is nearly identical to the UR for the former but not for the latter. This point is not too useful because a variety of things are learned in any situation. The classical defense experiment is perhaps the best example, wherein a leg-flex response is learned as well as perhaps autonomic responses. Therefore, the CR-UR similarity greatly depends upon which facet of the experiment is being scrutinized by the experimenter. Stated in more general terms, stimuli do have unconditioned effects, even weak stimuli like lights which produce such responses as pupil contraction. If the experimenter wishes to investigate those effects then one is confined to them by the choice of the stimulus (a classical experiment). However, the experimenter may wish to study other, more arbitrary, responses even though the chosen US produces a UR, in which case there is no confinement and the experiment becomes an instrumental one. In other words, if a reflexive-type response is studied, there must be a limitation to the choice of the US. If other arbitrary responses are the focus, no restriction is present. The converse is true as well: Once a US is chosen, the CR is necessarily restricted in classical but not instrumental conditioning. The resemblance of the CR to the UR then is not a valid basis for distinguishing classical from instrumental conditioning. Rather, it more simply reflects the freedoms and limitations in each type of experiment for choosing which facet of the organism's behavior is to be investigated.

In summary, differences in the response characteristics cannot be used to distinguish classical from instrumental experiments and, hence, do not suggest separate acquisition processes. Responses, varying in reflexiveness, may be conditioned by either method, although it is often true that a limitation is imposed on the range of possible CR's once the US is specified in classical conditioning.

Reinforcement Class

A third area for distinguishing classical from instrumental involves the nature of the US or reinforcer. If there were a pure Pavlovian US that could not function as a reinforcer in instrumental conditioning, then it might strongly suggest the existence of separate acquisition processes. Specifying such pure stimuli, however, is not an easy task. It is clear that many common stimuli may be used as US's in both types of learning experiments, e.g., food, shock.

As mentioned in Chapter 2, the sensory preconditioning experiment might represent the minimum conditions for Pavlovian conditioning. Therefore, it is possible that this phenomenon demonstrates classical but not instrumental conditioning insofar as the US is sufficient for establishing a classical CS-US association but not for maintaining an instrumental response. The US may be uniquely Pavlovian, because it is sufficient for classical but not instrumental conditioning.

There is a counterargument to the above, namely that numerous experiments have demonstrated that mild sensory stimulation, like a light, may be used as an instrumental reinforcer. Stimuli which are viable in the sensory preconditioning situation will also reinforce instrumental behavior (see Eisenberger, 1972, and Glanzer, 1958, for reviews). For example, Kish (1955) placed mice in a totally dark lever-box for 25 minutes each day for seven days. On day eight, a dim light was presented for 0.5 seconds to the experimental group for each lever-press, but not to the control group. Lever-pressing increased dramatically for the experimental subjects but not for the controls. This study clearly illustrates that such stimuli may act as instrumental reinforcers. Therefore, they may not be pure Pavlovian reinforcers since their effectiveness is not limited to the sensory preconditioning situations.

There is one problem with this counterargument, however. Kish (1955) stated that the conditioning effects he observed were enhanced by (required?) a long habituation period of seven days. In fact, it is now clear that the effect of many of these nonbiological reinforcers may be weak and/or quite transient (Barnes & Kish, 1961) or not effective at all (Symmes & Leaton, 1962). Furthermore, they usually are related to prior sensory deprivation (Fowler, 1967; Eisenberger, 1972). Therefore, the strength of these stimuli as instrumental reinforcers, at least for subjects that have not undergone sensory deprivation, is quite marginal.

In conclusion, the evidence is not strongly in favor of the two-process interpretation. Weak stimuli may be used classically (sensory preconditioning) as well as in instrumental experiments, although special methods may be required to show this effect, e.g., sensory deprivation.

A second reinforcement class in Pavlovian conditioning which may be distinct involves the US's in interoceptive conditioning. According to Rescorla and Solomon (1967), these stimuli would probably go unnoticed in an instrumental experiment and thus would fail to act as reinforcers. As unique US's in classical conditioning, however, they are more promising than the neutral US's in sensory preconditioning. Very little research has been done on the problem.

An example of an interoceptive stimulus is direct stimulation of the brain, rather than the viscera, which will serve as a US in classical but not instrumental conditioning. Doty and Giurgea (1961) conditioned a limb-flex by directly stimulating the motor cortex (US), but later demonstrated that the same brain stimulation was unable to serve as an instrumental reinforcer. However, more recent experiments have not shown such a dichotomy when using other types of brain stimulation. For example, Malmo (1965) conditioned a heart-rate decrease using stimulation of the brain as the US and found that, on a later test, the same stimulation served as an instrumental reinforcer for a lever-press response. Therefore, the evidence claiming that brain stimulation, or interoceptive US's in general, is a unique Pavlovian US is promising but equivocal at this time.

In summary, the general evidence for the two-process position, in considering both response and reinforcement class, is not very compelling. The converse statement seems equally true—there is little evidence which demands a one-factor interpretation.

What is one to believe? There are several ways that psychologists have chosen to cope with this ambiguity. The first, and perhaps most appropriate, has been to recognize the complexity of the problem and continue to search for new perspectives. A second has been to step back and reopen the question of what is learned. If that question can be answered, new principles may be outlined which do not involve analysis of classical versus instrumental acquisition processes. The third, as mentioned previously, has been to state that the question of whether there is one or two acquisition processes need not be asked. Here, an analysis of the functional relationships between the operations of the experimenter and the systematic and observable responses of the subject is the important focus for inquiry, not the theoretical nature of the underlying processes.

The first solution is best represented by the recent work of Solomon and his associates, which is considered below. The second is seen in the work of Bolles (1972) and, along with the third, will be considered in more detail in Chapter 10. Clearly, these three approaches are not mutually exclusive.

Analysis of Classical and Instrumental Interaction

The two-process position, in addition to stating that classical and instrumental conditioning represent separate and unique acquisition processes, postulates that the two learning situations interact. More specifically, the theory has claimed that Pavlovian conditioning mediates or influences instrumental behavior. Therefore, a different approach to the question of two-process theory is to explicitly investigate the joint action of classical and instrumental experiments, rather than attempt to separate them. The theoretical importance of these interaction studies has been emphasized by Solomon (see Rescorla & Solomon, 1967), whose studies suggest a basis for the two-process theory to the extent that Pavlovian processes mediate instrumental behavior.

Conditioned Emotional States

It was mentioned in Chapter 2 that two responses were conditioned in a classical defense experiment: a motor response like a leg-flex, and anxiety. There is reason to believe that all Pavlovian situations involve a form of conditioned emotion.

Conditioned emotional states were articulated by Mowrer (1960), who claimed that food and other primary appetitive rewards produced a basic, unconditioned effect, namely a state of pleasure. On the other hand, aversive stimuli like shock produced pain. This clearly is a hedonistic system, pleasure producing approach and pain eliciting withdrawal. Through Pavlovian conditioning, however, neutral stimuli acquire the capacity to elicit these emotional states as indicated in Figure 4-1. Thus, a CS paired with shock (upper left of Figure 4-1) acquires fear-eliciting properties. A CS−, one signaling no shock or pain, comes to elicit a conditioned state of relief. The same is true of Pavlovian appetitive experiments. A signal for the presentation of an appetitive US produces a conditioned emotion labeled hope, while a CS−, which signals no food, elicits a state of disappointment.

In other words, during Pavlovian conditioning, the US produces not only a specific visceral or motor response (like salivation, heart-rate change, or paw-flex) but also a diffuse, emotional state along a pleasure-pain dimension. The association between the CS and US also involves the emotional state. Excitatory conditioning produces hope or fear (depending on whether the US is appetitive or aversive), while inhibitory conditioning (CS− followed by nonpresentation of US) leads to the opposite emotional states, disappointment or relief.

Interaction Studies

The interaction experiments have borrowed heavily from Mowrer's conception of conditioned emotional states. In studying the interaction between classical and instrumental processes, the experimental strategy has been to manipulate the Pavlovian (emotional) state during instrumental responding. The change in instrumental behavior

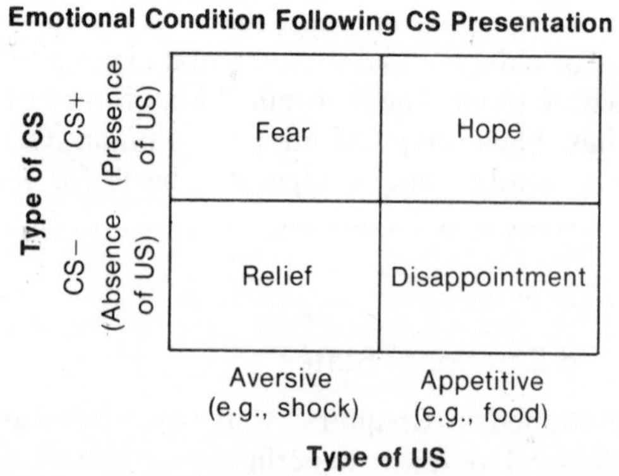

Emotional Condition Following CS Presentation

Type of CS	Aversive (e.g., shock)	Appetitive (e.g., food)
CS+ (Presence of US)	Fear	Hope
CS− (Absence of US)	Relief	Disappointment

Type of US

Figure 4-1. Matrix illustrating the four basic conditioned emotional states.

reflects the influence (or mediation) of the underlying Pavlovian state.

The most notable example (mentioned in Chapter 2 in connection with Kamin's blocking experiment, p. 39) is the conditioned emotional response (CER) experiment. This technique, first described by Estes and Skinner (1941), varies somewhat from study to study. Usually, subjects are taught to lever-press for food. Interspersed are trials in which a CS is presented for up to 3 minutes and is followed by a brief shock. Subsequently, when the stimulus is again presented, eliciting a fear response, the subject's responding is interrupted. Furthermore, the degree of interruption—the extent of decrease in the pressing rate—becomes an accurate index of the amount of fear the subject has toward the stimulus. If the subject is very fearful, almost complete suppression is shown, whereas if little fear is produced, responding is relatively unaffected. In Mowrer's terms, the basic CER experiment represents the "intrusion" of a counteractive emotion (fear) on the existing, basically pleasurable state which normally mediates appetitive conditioning.

An example of the power of the CER technique for studying fear is an experiment by Annau and Kamin (1961). The variable which they manipulated was shock intensity. During acquisition, an intensity of .28 mA produced virtually no suppression of responding. The CS followed by .49 mA, however, produced a response rate only about one half that of the baseline rate just prior to the CS presentation, while suppression was virtually complete for intensities of .85, 1.55, or 2.91 mA. This finding is shown in the left panel of Figure 4-2. The suppression ratio is .5 when no change in rate is observed, but approaches zero as suppression becomes total. According to the acquisition data, only three levels of fear were demonstrated; the three higher intensity groups all showed total suppression. However, as shown in the right panel of Figure 4-2, differential fear was demonstrated during extinction (when the CS was presented without the shock). The group receiving the highest intensity took the longest to recover; the group showed no suppression in response to the CS only after about nine days. In other groups, where fear was less intense, recovery was quicker. In summary, the CER technique has come to be a valuable tool in measuring fear. In more general terms, it demonstrates the interaction of emotional states, or, more precisely, the influence of a superimposed Pavlovian fear state on ongoing instrumental behavior.

As pointed out by Rescorla and Solomon (1967), there are other possible ways of presenting Pavlovian stimuli in instrumental situations, many of which are represented in Figure 4-3. The rows identify Pavlovian stimuli explicitly paired or not paired with a US (CS+ and CS− respectively) for either appetitive or aversive situations. The conditioned emotional states are also indicated. The columns represent

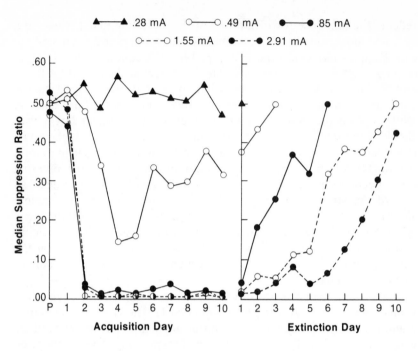

Figure 4–2. Median suppression ratio as a function of shock intensity during acquisition (left panel) and extinction (right panel).

the instrumental situation on which the classical stimuli are superimposed. The arrow in each cell predicts the effect of the Pavlovian stimulus on the instrumental behavior as either suppressive (↓) or facilitative (↑). Therefore, cell 5 corresponds to the CER experiment described above (a stimulus paired with an aversive US is presented during instrumental appetitive conditioning and the instrumental behavior is suppressed).

If fear is a counteractive force in an instrumental appetitive experiment (CER), then it should do the opposite in an instrumental aversive experiment—it should facilitate responding rather than suppress it. That is, an aversive CS+, when superimposed on, say, avoidance responding, should augment the emotional state and produce an increase in performance (cell 6, Figure 4-3). There are now a great many experiments which support this prediction. For example, Martin and Riess (1969) trained rats to press a lever which postponed shock. By responding at a stable rate, shock could be avoided altogether. Next, the subjects were given classical conditioning with either a .25, .58, 1.9, or 4.9 mA shock (US). The light (CS) was then superimposed during a subsequent avoidance session. The results,

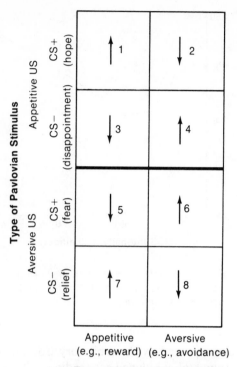

Figure 4–3. Matrix illustrating the interactions between Pavlovian stimuli (emotional state) and instrumental responding. Arrows indicate whether the superimposed Pavlovian stimulus facilitates (↑) or inhibits (↓) the instrumental task.

shown in Figure 4-4, clearly illustrate that the rate increased during the fear CS; a ratio of .5 indicates no effect while higher ratios indicate an increase in the avoidance response rate. Like Annau and Kamin's (1961) CER study (except in the opposite direction), the effect was magnified by greater fear (shock intensity). Scobie (1972) has recently shown that the increase in rate may depend not simply upon the US intensity but rather upon its strength relative to the intensity of the avoidance shock. Regardless, the augmentation of fear by superimposing an aversive CS+ interacts with the instrumental avoidance response.

Numerous studies have also presented an aversive CS− to the subjects during avoidance conditioning (cell 8, Figure 4-3). For example, Rescorla and LoLordo (1965) showed that such a procedure reduced the rate of responding, suggesting inhibition of fear by the CS−. In other words, the underlying fear state which mediates

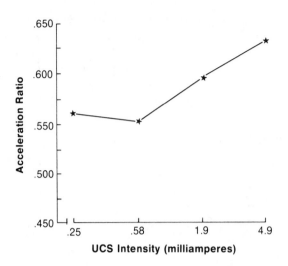

Figure 4–4. Mean acceleration ratio as a function of shock intensity. A ratio of .5 indicates no change, while higher ratios indicate an increase in avoidance responding rate.

avoidance was counteracted by the inhibitory CS−, the result being a reduction in instrumental avoidance responding.

Referring back to the CER technique, the information on CS− presentation during appetitive behavior (cell 7, Figure 4-3) is equivocal. The predicted effect (enhancement of response rate) has been demonstrated by Hammond (1966), but only when the baseline responding rate was below normal.

The experiments described above indicate the effect of fear and relief signals on instrumental responding. Converse situations have also been investigated: Superimposing appetitive stimuli on instrumental behavior. Although intuitively less obvious, the Pavlovian stimuli presumably establish conditioned emotional states of hope or disappointment which in turn influence instrumental behavior according to the predictions shown in Figure 4-3.

Examples of instrumental appetitive behavior being enhanced (or depressed) by classical appetitive stimuli are provided by Bolles, Grossen, Hargrave, and Duncan (1970) and Trapold and Winokur (1967). Trapold and Winokur measured lever-pressing rates, while Bolles et al. (1970) measured speed of running in an alley during the presentation of the classical stimuli. The predicted effects of appetitive CS+ and CS− presentations on rate of responding or running speed were observed in both experiments, although Bolles et al. (1970) found them only in extinction. The CS+ enhanced responding (cell 1 of

Figure 4-3) while the CS− inhibited it (cell 3).

An interesting finding is that improvement in discrimination learning may be achieved by prior classical conditioning. Mellgren and Ost (1969) first trained rats to press a lever on a VI schedule for water reward. Pavlovian conditioning was then given (the lever was removed from the cage), during which a tone (CS+) was followed by water, and a light (CS−) was followed by no reward. A second control group received random water presentations without regard to the tones and lights. In the test phase, the lever was reinserted and all the subjects were allowed to obtain water during the presence of the tone but not the light. Prior conditioning facilitated acquisition for the experimental group but not the control group.

Finally, there are numerous experiments which demonstrate effects of classical appetitive stimuli on instrumental aversive behavior. Again, the predictions (cells 2 and 4 of Figure 4-3) have been basically supported, suggesting that the appetitive and aversive emotional states are counteractive (e.g., Bull, 1970; Davis & Kreuter, 1972). For example, in an experiment by Grossen, Kostansek, and Bolles (1969), rats could postpone shock by running from one side of the shock box to the other and back again. A stable rate of responding was established before each subject received classical appetitive training. One group got tone-food pairings, a second tone-no food, and a third group received randomly presented tone and food. In the test phase, the tone was presented to all groups during avoidance performance, and responding rate, as a percentage of the pretone baseline rate, was measured. As shown in Figure 4-5, the conditioned CS− enhanced avoidance performance (avoidance rate went up), while the CS+ decreased performance (avoidance rate went down). The control group, in contrast, was not affected by the tone.

In summary, the interaction studies have illustrated the effects of combining conditioned emotional states—of superimposing either a compatible or counteractive Pavlovian state on ongoing instrumental behavior. The implication is that the underlying state which mediates the instrumental behavior is subject to specific laws of classical conditioning. The precise involvement of Pavlovian phenomena in instrumental behavior (beyond simply showing interaction effects) may be seen in an experiment by Rescorla (1967a).

Rescorla elicited a reduction of avoidance responding based not upon an inhibitory CS− (paired explicitly with no shock), but on inhibition of delay stemming from a long-duration CS+. Dogs were first trained to avoid shock until a stable rate of responding had been established. One group was then given classical conditioning with a 30-second CS-US interval; the controls received the same treatment except that the tone (CS) and shock (US) presentations were ran-

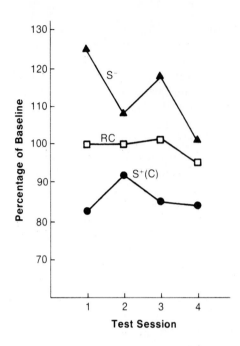

Figure 4–5. Mean percentage of avoidance responding baseline on the test sessions as a function of a Pavlovian appetitive CS+ presentation [group S+(C)], CS− presentation (group S−), or a noncontingent stimulus presentation (group RC).

domized. After sufficient training the subjects were tested by presenting the classical CS+ during the avoidance behavior.

Figure 4-6 illustrates the results. During the seconds prior to the CS presentation, mean rate of responding was the same in both groups. During the CS, however, the rate dropped significantly for the experimental subjects at first, but toward the end of the 30-second interval it accelerated beyond the control level. In contrast, the control response rate fluctuated but was not greatly affected by the tone presentation. In summary, then, the avoidance behavior was first inhibited due to inhibition of delay, but later in the interval it was facilitated. Rescorla's elegant study is evidence of Pavlovian control of instrumental behavior and suggests that classically conditioned states underlie instrumental behavior, since even specific Pavlovian phenomena, such as inhibition of delay, may be detected.

In conclusion, the interaction studies have generally supported a two-process interpretation of learning insofar as classical associations influence instrumental responding. Although the exact nature of this influence is not known, it may be in the form of an underlying

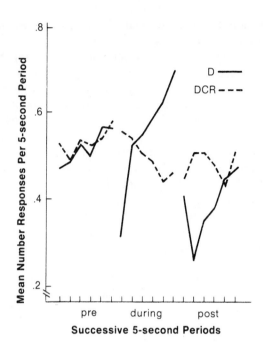

Figure 4–6. Mean number of avoidance responses per 5 seconds prior to, during, and following the presentation of the aversive CS+.

emotional state or the interaction of reward expectancies. Regardless, the interaction studies have provided a new focus and technique for assessing the concurrent action of classical and instrumental operations.

Are there two types of learning or one? The answer to that question still requires more study. However, it is clear that the two processes interact in a predictable fashion. This success in showing involvement of classical procedures in instrumental behavior suggested to Rescorla and Solomon (1967) that ". . . the version of two-process theory postulating that the concomitance we do observe between CR's and instrumental responding is mediated by a common central state, and the changes in that state are subject to the laws of Pavlovian conditioning [p. 178]."

Summary

Classical and instrumental conditioning may represent not only different training techniques but different underlying learning processes. One possible way of distinguishing classical from instrumental learning is by identifying conditionable responses which may be unique to each process. Historically, classical conditioning has involved autonomic responses, while instrumental conditioning has involved skeletal motor responses. This distinction, as well as one between voluntary and reflexive response classes, is a questionable one. The distinction between classical and instrumental conditioning on the basis of unique reinforcers is not compelling either. Most US's can be used as instrumental reinforcers although there are exceptions.

Given the complexity and ambiguity of separating these two processes, a different approach has been taken: the investigation of classical and instrumental interaction. These experiments superimpose a classically conditioned emotional state upon ongoing instrumental behavior. The change in instrumental responding, then, reflects the nature of the classical mediation. One clear example is a conditioned emotional response experiment in which a Pavlovian fear CS disrupts ongoing instrumental appetitive responding. Such a fear CS, however, augments avoidance responding while Pavlovian appetitive CS's enhance instrumental appetitive responding but decrease avoidance responding.

Avoidance

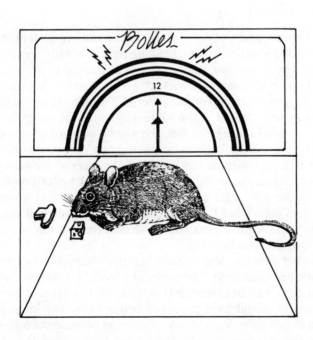

Introduction

Brogden, Lipman, and Culler (1938) performed an experiment in which the classical and instrumental techniques were explicitly compared. In their study, two groups of guinea pigs were trained to make a motor response in a small running-wheel to which electric shock could be applied. One group was given a buzzer (CS) followed by shock. The second group was treated in the same way but if the motor response (running) occurred during the CS, the shock was not presented and thus could be avoided. Initially, neither group performed the conditioned running response to the sound of the buzzer. However, with continued training, the avoidance group showed better conditioning (running to the CS only) as compared to the classical group. The experimenters concluded from these data that instrumental avoidance conditioning was a far superior method of training a motor response than was classical conditioning.

It was not entirely clear to early learning theorists why this result should have been obtained. The prevailing model for motor conditioning at that time was Bekhterev's (1913) classical defense experiment. That technique should have produced better motor conditioning than one where the shock was not always presented because the omission of shock for the avoidance group should have weakened the classical association. From the Pavlovian viewpoint, the basic problem then was to explain how the nonoccurrence of the US could reinforce the motor response to such an extent that the avoidance group would show superior conditioning.

The source of reinforcement was also a problem from an instrumental conditioning point of view. It was easy to understand escape from shock, but in the Brogden et al. (1938) experiment, shock was not presented on the avoidance trials, therefore no escape was possible.

From either viewpoint then, performance of the avoidance group should have been inferior. On the one hand, that group had fewer CS-US presentations. On the other hand, from the instrumental conditioning side, there was no obvious reason why the subjects should have responded if the shock was not being presented.

Acquired Drives

To understand the problem of avoidance, it is necessary to digress briefly and consider the notion of acquired drives. It was pointed out in Chapters 2 and 4 that conditioned emotional states are established during classical conditioning. The emotion of fear, as a consequence of classical defense conditioning, is a particularly obvious example. It is reasonable to infer that, due to the CS-US pairings, the subjects in the Brogden et al. (1938) experiment had developed fear.

The next step is to consider the possible role of fear for the avoidance group. A classic study by Miller (1948) illustrated the motivational properties of fear. Miller placed rats in the black side of a two-compartment box, then presented shock and allowed the animal to escape through a door into the other (white) side of the box. This procedure was repeated a number of times. At a later time, he tested the animal by placing it in the black side (without administering shock) and allowing it to run to the white side. Miller found that the animals learned to run into the white side without shock being presented. He also showed that the animals would press a small lever in order to open the door and gain access to the white side. Presumably, the color black, as well as other cues in the black side of the cage, had acquired aversive CS properties—they elicited conditioned fear. The important point, however, is that the acquired fear was sufficient to motivate the animal's learning (e.g., lever-press).

Miller's experiment is one of a number of studies which clearly showed that fear is learnable (the subjects did not initially fear the black side), and that it can motivate the learning of new behavior in the absence of primary drive (shock). In Miller's experiment, the subjects learned to escape into the white compartment by pressing the lever not because they were experiencing primary drive (i.e., pain from the shock), but because they were motivated by fear.

One additional point should be noted: The acquired fear is subject to the principles of classical conditioning, as has been demonstrated in a variety of ways. For example, Mowrer and Aiken (1954) first trained rats to press a lever for food on a VI schedule. Next, they gave classical light-shock pairings but varied the onset of the CS relative to the US. For group 1, the CS occurred just prior to the US (normal forward conditioning). For the other groups, however, the CS was presented either simultaneously with the US onset, with the US offset, or following the US termination (for groups 2, 3, and 4 respectively). The subjects were subsequently replaced in the lever-cage and presented with the fear CS following each lever-press. The extent of suppression of lever responding was taken as an index of fear (i.e., a conditioned emotional response experiment). As shown in Figure 5-1, the groups did not differ for the first five minutes, when each lever response was reinforced with food as usual. However, the degree of suppression, starting in the sixth minute, upon receiving the light rather than food, clearly shows that the forward procedure (group 1) had produced the most fear. The other groups showed only slight suppression of behavior in response to the light. Mowrer and Aiken concluded that contiguity (CS-US onset) was the important feature determining the strength of acquired fear; thus, fear is acquired according to classical conditioning principles.

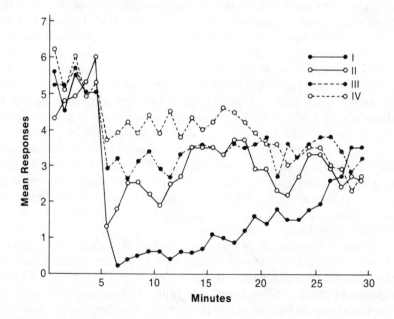

Figure 5–1. Mean number of responses for the different groups as a function of the minutes on the test.

Two-Factor Theory of Avoidance

It is clear from the studies discussed above that fear provides the motivation for avoidance responding and that fear is a classically conditioned component. There is a second component, however, involving the instrumental motor response. Given that fear motivated the response, what was the reinforcer?

In his famous paper, Mowrer (1947) identified both the classical and instrumental components in the avoidance response sequence. He stated that the acquired fear of the CS was learned according to Pavlovian principles, and the locomotor response was learned according to instrumental principles, the reinforcement being drive (fear) reduction. Mowrer assumed that the motor response was first made to escape shock (shock offset was the reinforcer) but then moved forward in time such that on subsequent trials the animal would execute an avoidance response which was motivated by acquired fear and was maintained, or reinforced, by fear reduction. In summary, this theory adequately accounted for the available data as well as the theoretical problem (brought out in the Brogden et al. study) of identifying the source of motivation (fear) and the source of reinforcement (fear reduction) for the avoidance response.

An important modification of the two-factor theory was made by Schoenfeld (1950), who argued that fear reduction was not the correct or parsimonious explanation for avoidance behavior. He postulated that once the CS acquired aversive properties during the early trials, the subjects would simply escape the noxious, internal cues which were elicited by the CS. Schoenfeld's point was subtle but important. For Mowrer, the reinforcement for avoidance was fear reduction, a reduction in a Pavlovian emotional state. In contrast, for Schoenfeld, the reinforcement for avoidance was CS offset, more simply the escape from or termination of the noxious CS.

Schoenfeld's position was an improvement over Mowrer's formulation mainly because the reinforcer was better specified. For Schoenfeld the locus of reinforcement was an event (CS offset) which could be isolated and examined. Schoenfeld's position emphasized the two sets of operations or two factors—fear acquisition and the motor response—and thus the duality of avoidance learning. The poorly specified fear reduction explanation of Mowrer was replaced by escape from the CS. This event was acceptable to the drive theorists as the locus of reinforcement, since it was the Pavlovian CS that elicited fear, and thus its offset should indicate the point of fear reduction. Once the reinforcer was identified (CS offset), subsequent research focused on that event, making avoidance conditioning the testing ground for two-factor theory.

Research focusing on the CS offset provided strong support for the two-factor theory. Much of it was done by Kamin (1954, 1956, 1957a,b). In one study, Kamin (1957a) reasoned that if the CS offset was the reinforcing event for avoidance, then a delay in that reinforcer should produce inferior avoidance learning, since it is well known that reinforcement delays do retard acquisition. Kamin ran four groups of rats in a shuttle box (where the response is running from one of two compartments to the other, and, on the next trial, back again) with a buzzer (CS). The buzzer came on 5 seconds prior to the shock and terminated following a response and a delay interval. Group 1 received no delay of CS offset for both an escape and avoidance response; the CS was turned off immediately following a response. However, for groups 2, 3, and 4 the buzzer offset was delayed either 2.5, 5, or 10 seconds, respectively.

The results are shown in Figure 5-2; percent avoidances for each block of 10 trials is given for each group. It is clear that the best learning occurred in the no-delay group and that performance varied

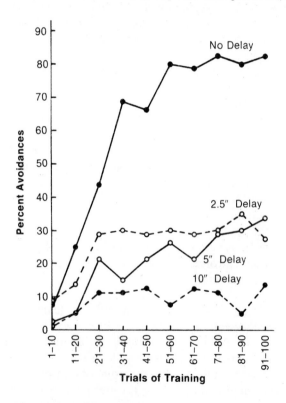

Figure 5–2. Percentage avoidance responses during training as a function of the delay of CS offset.

inversely with CS offset delay. The important point of this study is that it strongly suggests that the CS offset is the reinforcer for avoidance, since, as is usual, delaying that event resulted in poorer performance. Consequently, the study supports the two-factor theory in that fear reduction, which occurs when the fear CS is terminated, is the reinforcer for the instrumental avoidance response.

In another shuttle-box study, Kamin (1956) had four groups of rats. The first was a regular avoidance group; on avoidance trials, a response terminated the CS as well as avoided the US. A second group, labeled the Pavlovian group, could neither terminate the CS nor avoid the US. This group received classical CS-US pairings. The third group of animals (avoid-US group) could avoid the shock by making the appropriate shuttle response within 5 seconds after the CS came on but could not terminate the CS, i.e., the CS continued for the full 5 seconds. Therefore, the CS offset was delayed. For the last group (terminate-CS group), an avoidance response terminated the CS but did not eliminate the occurrence of the shock (US).

As shown in Figure 5-3, Kamin found that the normal avoidance group learned the best; performance was worst in the classical group (essentially replicating the findings of Brogden, Lipman, & Culler, 1938); and the other two groups were intermediate. According to

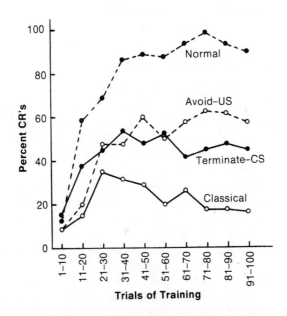

Figure 5-3. Mean percentage avoidance responses during training as a function of the ability to terminate the CS and/or avoid US.

Kamin, the terminate-CS group was inferior to the normal group because their avoidance responses were punished by the onset of shock even though they had been able to turn the CS off. The avoid-US group was inferior because the CS offset was delayed (as in his study discussed earlier, Kamin, 1957a). In summary, these data supported the two-factor theory insofar as the CS offset was shown to be important for avoidance learning. Coupled with other experiments (e.g., Kamin, 1954; Mowrer & Lamoreaux, 1942), Kamin's study in general demonstrated the importance of response-CS offset synchrony, and suggested that the CS offset acts as the reinforcer for the instrumental avoidance response.

Challenges to the Two-Factor Theory

On the surface, the two-factor theory in general, and the escape-from-CS hypothesis in particular, appeared to be a viable and comprehensive account of avoidance responding. The theory was elegant in specifying the two components involved in avoidance learning, and the data supported the basic predictions, e.g., the effect of delaying the CS offset. However, a great deal of evidence has recently accumulated strongly suggesting that at least the escape hypothesis—the strong, or specific, formulation of the two-factor theory—is quite wrong (see Bolles, 1967, and Herrnstein, 1969, for reviews).

Discriminative Stimulus Functions

The first area of evidence deals with the discriminative or signaling capacity of the CS. Sidman (1953a,b) was the first to show that an animal could effectively avoid or postpone a shock even though the shock was not preceded by an external CS. In this type of avoidance responding situation (called Sidman or nondiscriminated avoidance) all shocks may be avoided if the rate of pressing is sufficiently high.

At first glance, Sidman's finding raises a problem for the escape hypothesis: Because there was no CS, there was no CS offset (i.e., reinforcement), yet there clearly was avoidance performance. To explain why avoidance occurred in the apparent absence of reinforcement, Sidman claimed that all behavior other than the correct avoidance response was punished by shock. The avoidance response per se was at least more removed in time from shock punishment and, in fact, was reinforced when it terminated the aversive cues associated with the other punished behavior. In other words, the avoidance response was not severely punished because shock was postponed by it; it also terminated punished (aversive) behavior and therefore was reinforced.

A simpler explanation was offered by Anger (1963). He argued that the subjects developed a sense of the time interval between shocks, and, as a consequence, the temporal interval itself functioned as a CS. During the interval, internal, aversive stimuli would build up to a maximum; an avoidance response was then reinforced because it eliminated those aversive stimuli. Anger's contribution was important for the two-factor viewpoint because it pointed out that the CS need not be an external signal, such as a buzzer; rather, even internal stimuli which occurred during the temporal interval could function as the CS and their offset as the reinforcer (see Gibbon, 1972, for a review).

Modifications of the Sidman design, however, have led to several other more serious problems for the two-factor theory. In two such studies, an external stimulus preceded the shock as usual. The subjects were allowed to press a lever (Sidman, 1955) or run in a wheel (Keehn, 1959) at any time to postpone not only shock but also the cue presentation. The outcome was that the subjects did postpone the shock but not the cue; they waited until the cue came on, and then responded to avoid the shock.

The problem is that if the CS elicited fear, then given the opportunity, the subjects should have responded in order to postpone that as well. Rather, the subjects appeared as if they were using the cue simply as a discriminative stimulus, one signaling the appropriate time for a response. These results suggest that the function of the CS is to inform the subject when to respond. The function of the CS offset, rather than reinforcing, more simply provides information that the appropriate response has been made—information which the subject would not otherwise have on an avoidance trial.

The notion that the CS offset is simply an informational cue has been confirmed by Bower, Starr, and Lazarovitz (1965). In this experiment, two results were obtained, both of which suggested that the CS offset functions as a discriminative signal rather than as a reinforcer. First, avoidance performance improved as a function of the degree of change in the noise (CS) following the response, regardless of whether the change was away from or in the direction of a baseline noise.

A more important result (also found by Bolles & Grossen, 1969, and D'Amato, Fazzaro, & Etkin, 1968) involved three groups of rats. A response from animals in group 1 terminated the tone (regular avoidance). For group 2 a response enabled the rats to avoid the US, but delayed the CS offset for 8 seconds. Group 3 also received an 8-second CS-offset delay, but during that 8 seconds, a second stimulus (light) was presented.

As indicated in Figure 5-4, group 2 performed the worst (confirming the results of Kamin, 1957a,b) because the CS was delayed.

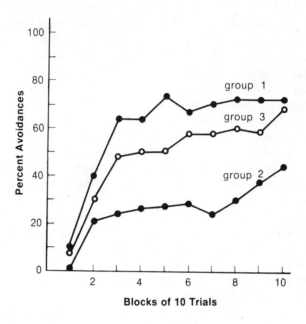

Figure 5–4. Mean percentage avoidance during training as a function of delay of CS offset (group 2), immediate CS offset (group 1), or CS offset delay with an additional light presentation (group 3).

However, the performance of group 3 was not inferior to that of group 1. Rather, the addition of a second cue during the delay period eliminated the deficit caused by the CS-offset delay. Bower et al. (1965) inferred that the cue had acted to inform the subject of its correct response in the absence of the usual signal, i.e., CS offset.

In summary, the above studies suggest that the CS functions as a discriminative stimulus. Its presentation elicits the avoidance response by indicating when the response should be executed. The CS offset functions to inform the subject that the correct response has been made; it is not the locus of reinforcement.

Additional Behavioral Studies

A number of additional behavioral studies have been conducted which also tend to refute the two-factor theory. The results in Kamin's (1956) study (see p. 95) have been interpreted in a different way by Solomon and Brush (1956). They expressed surprise that the avoid-US group learned better than the strict classical group (see Figure 5-3). Both groups received the identical CS duration, the only difference being that in one group, the avoid-US group, the US could be avoided.

This finding suggested to Solomon and Brush that the avoidance of the US, notwithstanding the effects of CS termination, is a source of reinforcement.

This point was pursued further in an elaborate study by Bolles, Stokes, and Younger (1966). In five separate experiments, the effects of CS offset, US avoidance, and US escape on avoidance performance were examined. By running in a wheel, four different groups of rats could either terminate the CS and avoid and escape the US (group TAE), terminate the CS and avoid the US (group TA), terminate the CS and escape the US (group TE), and only terminate the CS (group T).

As shown in Figure 5-5, the most important factor in determining avoidance responding was the ability to avoid the shock. Without that possibility, performance was very poor (groups T and TE). This study illustrated an important point: US avoidance may be the true reinforcer for the avoidance response. The CS offset does allow the subject to perform more efficiently but, in itself, is not a source of reward.

In further support of this idea, Black (1963) showed that subjects

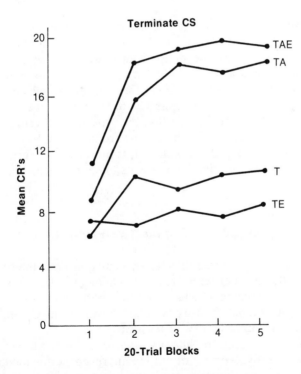

Figure 5–5. Mean number of avoidance responses during training as a function of the ability to terminate the CS (T) and/or escape the US (E) and/or avoid the US (A).

could learn avoidance responses even when using a trace procedure in which the CS was terminated well before the US onset, thereby leaving an empty interval. The CS offset could hardly be the source of reward for such an avoidance response, since it occurred prior to the response.

Finally, evidence against the two-factor theory was found by Herrnstein and Hineline (1966). A brief shock was administered according to two separate schedules. The frequency of shocks was random for both schedules, but the average frequency was less for schedule 1 than for schedule 2. If the subject simply sat in the cage and did not respond by pressing the lever, it received shocks according to schedule 1. If the subject did press the lever, the apparatus switched control over to schedule 2 and the next shock was delivered according to that schedule. Once the shock was delivered, control would revert back to schedule 1. Because schedule 2 administered a lower frequency of shocks, the average time between shocks was greater if the subject consistently responded. A response was more likely to produce a somewhat longer shock-free interval than no response, since the shock which would follow a response was determined by the lower frequency of schedule 2.

It is important to note that no external CS was presented, nor was there a fixed temporal CS since both schedules were randomized with respect to time. No escape or avoidance was possible; rather, the subjects simply had a choice between receiving a high frequency of shocks by not responding, or a lower frequency of shocks by pressing the lever. Herrnstein and Hineline (1966) found that the animals did learn to respond and concluded that the reinforcing event for avoidance, in general, must not be the escape from a fearful CS. Although the CS may have a signaling function, it is quite unnecessary since it is neither the source of motivation nor is its offset the locus of reinforcement.

Physiological Criticisms of the Two-Factor Theory

The research presented thus far strongly questions the role of the CS offset as the locus of reinforcement in avoidance learning. There is another class of arguments which focuses on the role of the CS as the motivator of avoidance. These physiological arguments generally don't support the two-factor theory, at least Mowrer's version.

In his formulation of the two-factor theory, Mowrer (1947) emphasized the involvement of autonomic responses as mediators of avoidance, claiming that the classically conditioned reactions which constituted anxiety were visceral in nature. Because it was long believed that the sympathetic nervous system functioned in times of stress, where the typical emotional response patterns were of a fight or flight nature, it was natural for Mowrer to postulate that the autonomic nervous system played a crucial role in avoidance behavior.

Two basic research strategies were adopted to investigate Mowrer's assumption. The first involved eliminating all, or most, of the sympathetic nervous system (surgically or chemically), and observing the subsequent effect on avoidance performance. Wynne and Solomon (1955) surgically sympathectomized dogs and tested avoidance acquisition and extinction in a shuttle box. They found greater variability in the behavior of the experimental subjects, and in some cases, slower acquisition scores. However, all the animals eventually did learn the task. Wynne and Solomon therefore concluded that the autonomic nervous system, while it did play some role in mediating avoidance (acquisition was retarded somewhat), was not the only physiological mechanism involved. Rather, hormonal, skeletal, and central nervous system factors were implicated as well. More recently, Van-Toller and Tarpy (1974) have also concluded that the autonomic nervous system has no essential or unique role in avoidance learning after examining many studies comparable to Wynne and Solomon's.

A second strategy was exemplified in a study by Black (1959). Rather than attempting to curtail the action of the autonomic nervous system and thereby limit fear, he concurrently measured heart rate (which is controlled by the autonomic nervous system) and avoidance. If the two are causally connected as Mowrer asserted, the physiological response should be highly correlated to the behavioral avoidance response.

Black required the subject to push a panel with its nose and thereby avoid a shock while, at the same time, he monitored heart rate. He found a significant relationship between heart-rate activity and avoidance performance during acquisition but not during extinction. That is, the heart-rate response to the CS extinguished more quickly than the avoidance response. Avoidance was being maintained in the absence of autonomic involvement, a finding which poses difficulties for Mowrer's theory.

In summary, these physiological studies were more directed at Mowrer's formulation that the autonomic nervous system was the only substrate of fear, than at the more general two-process position. These studies, thus, were important in showing that avoidance was not uniquely related to autonomic responding. Black (1959) summarized this point by saying ". . . it seems clear that theories such as Mowrer's, which simply treat autonomic behavior as an index of fear or anxiety without regard to other central and peripheral interrelations between autonomic and skeletal behavior are too simple to explain (these general) findings [p. 241]."

General Conclusions About the Two-Factor Theory

The result of these counterarguments has been to reject the strong version of the two-factor theory, which included two points that appear to be incorrect. First, the autonomic nervous system is not the exclusive substrate of fear as Mowrer believed, and therefore it is not a unique mediator of avoidance behavior. Rather, there are a variety of physiological systems which contribute to fear responding, only one of which is the autonomic system.

The second point involves the role of the CS, or more specifically, the CS offset. There is no question that the CS is capable of producing fear under certain circumstances (e.g., CER studies). However, it probably does not function simply as a fear-arousing stimulus during avoidance. Rather, it is an informative or discriminative cue which indicates to the subject the appropriate time for responding. Fear does not, in a unitary fashion at least, suddenly appear and disappear with the CS onset and offset. The CS offset, therefore, is not the locus of reinforcement but rather a source of information that the correct response has been made. Without the CS offset (e.g., when it is delayed) no such information is provided unless, of course, an additional cue is presented to the subject.

A modified version of the two-factor theory, however, appears warranted. This version more simply postulates the involvement of Pavlovian fear states in instrumental avoidance conditioning without specifying a unique reinforcing role for the CS offset. The interaction studies, mentioned in Chapter 4, show such involvement, although some learning theorists do not believe that the Pavlovian influence is necessarily one of emotion (Bolles, 1972).

There is fear during avoidance; classical stimuli do modify the avoidance behavior; and informative stimuli improve performance. The most fundamental contingency which produces avoidance is the ability to avoid per se rather than to terminate either the CS or shock. In contrast, there is not a unique or necessary autonomic involvement, nor is an explicit CS necessary for avoidance learning.

An important and radically different theory of avoidance learning has recently been advanced by Bolles (1970). He claimed that an animal in its natural habitat could not possibly avoid its predators by relying upon a gradual, trial-and-error learning process. A failure to learn such avoidance responses, of course, would mean death for the animal. More correctly, according to Bolles, the way in which an animal copes with aversive or dangerous situations is to make an innate defensive reaction which is specific to its species. Such species-specific defense reactions (e.g., flight, freezing, or pseudoaggressive responses) are not learned as the two-factor theory suggests but rather occur reflexively whenever the animal is aversively stimulated. In the laboratory avoidance situation, if the animal's first innate defensive reaction to shock is not effective in terminating or avoiding shock, the animal must abandon that response in favor of another species-specific defense reaction that works. Therefore, the avoidance contingencies restrict the range of appropriate behaviors to a narrow class of defense reactions, and defense responses that are not effective in meeting the avoidance contingency are suppressed.

Bolles' (1970) theory accounts for the fact that some avoidance responses are more easily learned by subjects than others. For example, a running response is easily conditioned in several trials, while a lever-press avoidance response is extremely hard, if not impossible, to condition (D'Amato & Schiff, 1964). This finding poses a problem for learning theory in general, since the reinforcement contingencies are the same in both instances, thus implying that the learning should not differ. Running, however, is an innate defense reaction and is therefore learned almost immediately, while lever-pressing is not part of the subject's natural survival behavior and is thus performed by the subject only after extended training in which more innate defense reactions such as running are suppressed.

The running versus lever-pressing comparison just discussed illustrates that species-specific defense reactions are easier to condition than nondefense reactions. However, an additional implication of Bolles' theory is that there is a hierarchy of defense reactions: Some defense reactions such as running are more dominant than others. The speed or ease of learning a particular avoidance response should reflect the relative dominance of that response in the hierarchy of defense reactions. This has been demonstrated in an excellent study by Grossen and Kelley (1972) in which rats could avoid a shock, one group by jumping onto a platform in the center of the cage and the other group by jumping onto one located at the edge of the cage. Previous research had indicated that rats naturally seek contact with objects (remain near the walls of the cage) when shocked, which suggests that jumping to a platform near the wall would be a more dominant defense reaction than jumping to one away from the wall. Grossen and Kelley did find that the group required to jump toward the wall learned more readily than the group required to jump toward the center, thus supporting the notion of a defense-reaction hierarchy.

Bolles' theory is a successful alternative to the two-factor theory, emphasizing the interaction of innate behavior patterns and environmental (avoidance) contingencies. Furthermore, his approach is typical of a broader trend in learning theory that considers principles of learning within the context of the evolutionary history of the species (see Chapter 10 for a more detailed discussion of this topic). The information value of the CS is certainly a viable factor in avoidance-learning experiments, as is the notion of Pavlovian fear-instrumental behavior interactions. Acknowledging the innate defense patterns, however, provides another dimension or focus to the situation and thus helps to develop a more comprehensive theory that accurately describes the mechanism for avoidance learning.

Factors in Avoidance Conditioning

Avoidance conditioning usually involves escape contingencies, at least early in training. However, in addition to the contingencies which influence escape behavior are those variables which influence the rate at which subjects learn to anticipate and thereby avoid the shock.

Intensity of US

The role of US intensity in avoidance behavior has received considerable attention primarily because of the implication that higher shock levels should produce greater fear and, therefore, better avoidance performance.

Early studies by Kimble (1955), Brush (1957), and Boren, Sidman, and Herrnstein (1959) indicated that performance, as measured by response speed and percentage of subjects that learn to avoid, was generally better with higher shock intensities. More recent studies, however, have questioned this generalization. In fact, Bolles and Warren (1965), D'Amato and Fazzaro (1966), Levine (1966), Moyer and Korn (1964), and Kurtz and Shafer (1967) have all shown that avoidance performance is inversely related to shock intensity. Higher intensities produce inferior avoidance.

For example, Moyer and Korn (1964) trained rats in a shuttle box to avoid shock which varied in intensity from .5 to 4.5 mA for separate groups. As shown in Figure 5-6, performance was worse with higher intensities. The authors hypothesized that stronger shock disrupted avoidance learning because competing responses, such as crouching or freezing, were elicited. Thus, with higher levels of fear due to higher shock intensities, these responses competed with the correct running response. The principle that avoidance performance decreases with higher shock intensities appears to be widely supported, although the reason is that competing responses disrupt the performance, not that fear is diminished at high-shock intensities.

This latter point has been confirmed in two interesting and separate ways. First, D'Amato, Fazzaro, and Etkin (1967) reasoned

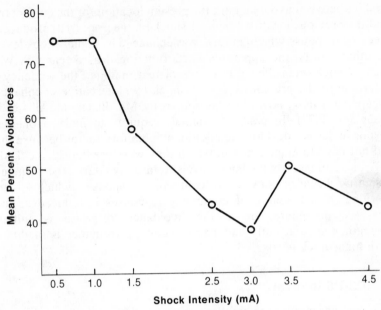

Figure 5–6. Mean percentage avoidance responses as a function of shock intensity.

that avoidance would be maintained at higher levels with higher shock intensities if the response was first acquired under a low intensity. They therefore trained rats to press a lever to avoid a weak shock, and once the response was adequately established, switched the intensity to a higher value. The performance improved greatly under those conditions, suggesting that the predicted positive relationship between performance and shock intensity is valid, provided that the avoidance response is initially strengthened while the competing responses of freezing and crouching are extinguished.

A second approach to this problem was taken by Thieos, Lynch, and Lowe (1966), who varied shock intensity under two separate conditions. One set of subjects was trained to make a shuttle-box (two-way) response, while the remaining subjects were trained on a one-way avoidance response, i.e., always running from a start box to a separate, discriminable goal box. Performance decreased with higher shock intensities for the two-way response but not for the one-way response. Thieos *et al.* (1966) concluded that the disruption at higher intensities for the two-way shuttle response was due to competing responses. However, they further identified the cause of these crouching or freezing responses as the tendency to not enter the compartment where shock had just been delivered on the preceding trial. The conflict between avoiding both the present location for the current trial and the previous location (locus of shock on the prior trial) produced a freezing response which, in turn, was enhanced by higher shock levels. In other words, the apparatus itself functioned as a fear CS. With higher shock intensities, this source of fear inhibited the tendency of returning to the previous locus of shock (or, alternatively, enhanced escape to a new, novel box as shown by McAllister, McAllister, & Douglass, 1971). In contrast, animals required to make a one-way response did not develop the additional inhibitory factor because they did not have to enter a box in which they were previously shocked.

In summary, high-shock levels retard avoidance performance primarily because they elicit competing responses which disrupt learning. If the source of competing responses is reduced as in a one-way procedure, or if the avoidance response is initially strengthened sufficiently under low shock, performance is facilitated with high-shock levels.

CS-US Interval

Another major variable that has received considerable attention is the CS-US interval, primarily because contiguity is an essential feature in classical fear conditioning. In general, however, the optimum CS-US interval for avoidance is much longer than what is usually found in classical defense conditioning, although this holds more for some species than others.

Low and Low (1962a) varied the CS-US interval for separate groups of rats and administered 100 avoidance acquisition trials in a shuttle box. The intervals used were either 2, 4, 6, 8, or 10 seconds from the onset of the CS to the onset of the US. The authors found that speed of responding decreased as a function of the CS-US interval—the longer the interval, the slower the response. However, as shown in Figure 5-7, number of avoidance responses or the number of trials to reach four consecutive avoidances was inversely related to interval length, i.e., the rate of avoidance acquisition improved as the time between the CS and US increased. Those authors concluded that there was more opportunity to make an avoidance response with longer CS-US intervals, and, hence, the performance was better.

In another study, Black (1963) varied the CS-US interval both in trace and delayed procedures. He found that delayed conditioning was superior to trace (confirming Kamin, 1954) and that, for delayed conditioning, performance was an inverted U function of the CS-US interval. That is, an increase in the interval from 5 to 10 seconds produced superior performance, whereas further increases to 30 seconds caused a decrement. Thus, the data from these two studies are

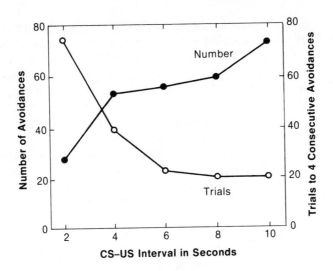

Figure 5–7. Number of avoidance responses (left ordinate) and trials to four consecutive avoidance responses (right ordinate) as a function of the CS–US interval.

complementary. Low and Low (1962a) studied intervals varying from 2 to 10 seconds and found performance positively related to the interval. Black found the same effect up to 10 seconds but went on to show that longer intervals produced a decrement. Therefore, the general inverted U shaped function appears to be valid although Bolles, Warren, & Ostrov (1966) did not find a decrement at the longer intervals for a lever-press avoidance. The cause for the decrement at longer intervals is not clearly known, although it is perhaps related to inhibition of delay (as discussed in Chapter 2).

It is interesting to note the varied data that have been obtained for different species. Behrend and Bitterman (1962) trained goldfish to avoid shock and found that performance improved (number of avoidance responses increased) as a function of the CS-US interval up to 60 seconds. Bitterman (1965) later noted that this function could be derived from the one obtained when general activity was classically conditioned, suggesting that avoidance in goldfish represents such a classically conditioned response. Experiments on dogs (Brush, Brush, & Solomon, 1955) and cats (Schrier, 1961) showed either small decrements or no decrement, respectively, in performance with longer intervals. In summary, these studies suggest that the CS-US interval is related to phylogenetic level. With a species low on the phylogenetic scale, such as fish, the correspondence between classical and instrumental conditioning functions is high, whereas the correspondence decreases somewhat for rats and considerably more for dogs and cats.

A final point concerns the variability of the CS-US interval. Low and Low (1962b) trained different groups of rats to avoid shock in a shuttle box. One group received a 6-second CS-US interval, while the other two groups received varied intervals with an average of 6 seconds. Number of avoidance responses in 100 trials did not differ between groups, but the speed of responding was positively related to the degree of variability—the more variable the interval, the faster the responding. Low and Low concluded that the lack of temporal consistency which should have retarded performance actually had the effect of emphasizing the CS onset as the salient cue for avoidance. Therefore, because the subjects in the variable-interval groups responded exclusively to that cue, the response speed was enhanced.

Intertrial Interval

A number of investigators have examined the effect of intertrial interval duration on avoidance conditioning. Brush (1962) trained rats to make a shuttle-box avoidance response. The intertrial interval was either .5, 1, 2, 5, 10, or 20 minutes. Figure 5-8 shows the results for the last 20 trials. The number of avoidance responses increased to a maximum at 5 minutes and then declined for the two longer intervals. Speed of response showed the same relationship except that the peak was at 1 minute. It thus appears that avoidance responding becomes

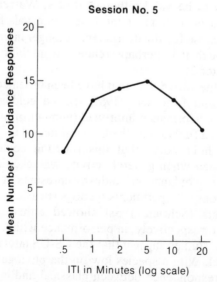

Session No. 5

Figure 5-8. Mean number of avoidance responses as a function of the intertrial interval on the fifth session of training.

both more probable and faster with longer intertrial interval durations until a maximum, after which responding again decreases. Similar results have been shown by Kurtz and Shafer (1967) and Levine and England (1960), although in those studies, the range of durations was considerably less.

Brush concluded that as the intertrial interval increased up to 5 minutes fear of the apparatus extinguished, since apparatus cues continued to exist in the absence of shock during the intertrial interval. This, in turn, allowed for competing responses, such as freezing, which are elicited by apparatus cues to extinguish and performance to improve. With further increases in the intertrial interval length, however, competing relaxation responses may have developed, thereby producing a decrement in performance.

A similar relaxation hypothesis was also offered by Reynierse, Weisman, and Denny (1963). They gave two groups of rats 60 avoidance trials per day for 6 consecutive days. The two groups differed only in the amount of time spent in the presence of the shock or nonshock cues during the intertrial interval, i.e., the goal box versus a separate neutral box. The group which spent relatively little of the intertrial interval time in the presence of the shock cues performed significantly better than the other group. The authors interpreted the results to mean that those subjects had learned competing relaxation responses in the neutral box which had interfered with the avoidance performance.

In summary, it appears that avoidance performance first improves with longer intertrial intervals but then declines with further increases. This curvelinear effect is probably related to changes in fear (elicited by apparatus cues), although it is not clear whether the decrement is due to a net reduction in fear or an increase in a specific type of competing response.

Retention of Avoidance

One final temporal variable is the retention interval or the time between avoidance training and a retention test. Moyer (1958) found that retention of avoidance generally did not differ between groups ranging in delay from 1 to 32 days; the only significant difference was between the two extremes of 1 and 32 days. More recently, Kirby (1963) studied avoidance retention as a function of age, and, in support of Moyer's overall findings, found that older subjects showed no deficit on a 50-day retention test, indicating that forgetting did not occur within that time. However, younger subjects, although they originally learned as well as the older ones, were unable to retain the response and showed forgetting after 25 days. The general principle that memory improves with age holds true for a wide variety of responses, including appetitive tasks (see Campbell, 1967, and Campbell & Spear, 1972, for reviews). In summary then, there is little forgetting of an avoidance response for at least 50 days except for subjects trained at a young age.

Another interesting effect related to avoidance retention over a shorter time period was first demonstrated by Kamin (1957c). After 25 original avoidance-learning trials in a shuttle box, separate groups were given retention intervals of 0, .5, 1, 6, 24 hours, or 19 days. After the retention interval, 25 additional relearning avoidance trials were administered. As shown in Figure 5-9 (see next page), Kamin found a U-shaped function with the lowest retention occurring at the one-hour interval. Kamin hypothesized that the decreasing portion of the U-shaped function (from 0 to 1 hour) was due to forgetting. The rise in the curve following one hour on the other hand, represented an incubation of the avoidance habit.

Numerous studies have appeared since Kamin's original demonstration. The work of Denny and his associates (Denny, 1958; Denny & Ditchman, 1962) confirmed the basic U-shaped function but concluded that the decrement in avoidance relearning was due to incubation, or increases, of anxiety rather than forgetting (see McAllister & McAllister, 1967, for a review). Following the initial avoidance, fear of the apparatus increased and, correspondingly, competing freezing responses were elicited which suppressed avoidance behavior. The apparatus fear dissipated after the hour rest interval, restoring the performance to its original level.

Denny's (1958) hypothesis is supported by the fact that subjects that remained in the apparatus for the retention interval (during which time apparatus fear presumably extinguished) did not show the decrement in avoidance relearning after 1 hour compared to control subjects that spent the interval in their home cages.

It is clear that fear training, as opposed to escape, shock-stress, or unpaired CS and US presentations, is the necessary and sufficient condition for producing the effect (Brush, Meyer, & Palmer, 1963; Brush, 1964). Also, the effect is not only limited to poorly learned avoidance responses (Singh, Sakellaris, & Brush, 1971).

More recently however, a number of experiments have suggested at least two alternative explanations (for a third explanation see McAllister & McAllister, 1967). One is that there is actually a decrease in fear after an intermediate retention interval rather than an increase as Denny hypothesized. The second is that there is a memory-retrieval failure rather than a change in the intensity of fear.

Regarding the first, certain drug manipulations will eliminate the avoidance decrement, e.g., administration of ACTH (Singh, Sakellaris,

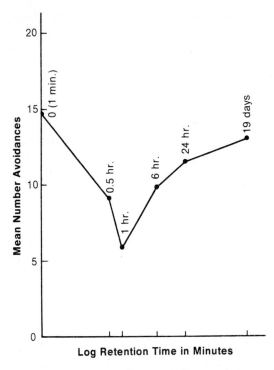

Figure 5-9. Mean number of avoidance responses as a function of the retention interval.

& Brush, 1971; Klein, 1972) or adrenalin (Hablitz & Braud, 1972). These drugs augment the internal, physiological conditions produced during fear, and when given prior to the retention test, presumably eliminate the relearning deficit because they restore the appropriate fear cues.

Other experimenters have proposed that the subjects actually forget the response after an intermediate retention interval (Bintz, 1970; Geller, Jarvik, & Robustelli, 1970; Klein & Spear, 1970). In the second experiment of the Klein and Spear study, rats were trained to avoid an electrified black box by remaining in an adjacent white box. Following a retention interval of either 5 minutes, 1, 4, or 24 hours for different groups, the subjects were required to learn the opposite response: actively avoiding shock in the white box by running to the black box. The intermediate groups learned this latter response much better than the 5-minute or 24-hour groups, and therefore concluded that memory of the original response was weaker in the intermediate groups. Because the intermediate groups forgot the first task, they had little trouble in acquiring the opposite response. These results fail to confirm an increase in fear (Denny's hypothesis), since crouching or freezing responses, which presumably would be enhanced by greater apparatus fear, would have hindered rather than helped the acquisition of the latter response.

In summary, the biphasic changes in behavior following aversive learning are well documented, although the exact cause is still unclear (see Brush, 1971, for a review). The effect is probably related to changes in fear, or more correctly, alterations of the internal physiological cues (Spear, Klein, & Riley, 1971), although it is not possible to separate the fear motivation and memory factors at this time (Singh, Sakellaris, & Brush, 1971).

Effects of Prior Shock on Avoidance

A final parameter deals with the effect of prior shock on avoidance learning. As discussed in Chapter 4 with reference to the interaction experiments, Pavlovian conditioned stimuli can modify ongoing avoidance behavior. However, if the classical fear-conditioning (inescapable shock) precedes the avoidance acquisition, the subjects are usually unable to learn the avoidance response as compared to subjects that are not given prior inescapable shock. In fact, they will not even escape the painful shock once it comes on. This effect was shown by Overmier and Seligman (1967) and Seligman and Maier (1967) and has been labeled "learned helplessness." The phenomenon is obtained only after inescapable shock is given, regardless of the place or intensity of shock, or whether or not it was preceded by a warning stimulus. The effect is also produced by escapable shock when the subject is required to withhold its response for several seconds following shock onset (Cohen, 1970). In contrast, prior shock, which is immediately escapable, has the predicted facilitating effect on later avoidance learning (Brush, 1970).

The cause for this effect is not clear. Maier, Seligman, and Solomon (1969) rejected explanations based upon either too little shock (which might produce adaptation) or too much shock (which might produce disorganization). Similarly, explanations based on the acquisition of responses during the shock which interfere with later avoidance learning seem inadequate, since Overmier and Seligman (1967) administered curare to their subjects during the classical phase of the experiment, and still observed the effect.

The learned-helplessness explanation contends that the subject first acquires the expectation that shock is delivered independently of its behavior. Then, during avoidance learning, this expectation transfers and prevents the subject from learning the avoidance contingency. Without prior shock treatment, no conflicting expectancy is developed.

Summary

In the investigation of avoidance conditioning, an early study showed that motor conditioning based on instrumental avoidance procedures was superior to that involving classical conditioning procedures. This finding raised the issues of motivation during the avoidance trial and the source of reinforcement for the avoidance response. There was no obvious reason why subjects should have responded as they did in the absence of shock. Later research suggested that fear was the source of motivation, and that such fear was acquired according to Pavlovian principles. Fear reduction, or more specifically, termination of the fear CS, was thought to be the source of reward. These two factors—classically conditioned fear to the CS and instrumental reinforcement produced by the CS offset—comprised the two-process explanation of avoidance learning.

Much research has supported the two-process theory. For example, if the CS offset was delayed following a response, performance deteriorated. However, more recent findings have challenged this position. The CS offset appears to function more as a discriminative signal. Avoidance of shock, notwithstanding CS offset, is the fundamental determinant of avoidance performance.

A number of variables affecting avoidance performance have been studied extensively. For example, performance improves with increasing shock intensity, but deteriorates above specifiable intensities. The decrement is probably due to freezing or crouching, which disrupts performance, since the decrement may be eliminated if precautions are taken to eliminate the competing responses. Similarly, performance improves but then declines with longer CS-US intervals or with longer intertrial intervals. Additional research has indicated that avoidance performance lessens after several hours but is restored to full strength by 24 hours. Several investigators have hypothesized that this short-term change is due to an increase in fear or memory-retrieval failure.

Punishment

Introduction

Punishment is defined as the operation in which a noxious stimulus, one that a subject will try to escape, is contingent upon the occurrence of a response. The operation is precisely the opposite from that used in reward training except that the response is followed by an aversive stimulus rather than an appetitive one. The general effect of punishment operations on behavior is suppression of the behavior—a reduction in response probability, or a decrease in vigor—although the explanation for this effect is not entirely clear.

Despite the fact that aversive stimuli are common features of everyday life, the systematic study of punishment in the laboratory has been relatively neglected. According to Solomon (1964), there are two basic reasons for this neglect. First, psychologists have avoided studying punishment because they have feared that neurosis or other undesirable side effects were an inevitable consequence of using aversive stimuli. The second reason is that some psychologists have believed that punishment was, quite simply, not effective in changing behavior. According to this latter view, the response suppression observed following a contingent aversive stimulus was only temporary, compared to the relatively permanent effects of positive reinforcement on behavior.

Solomon argued that both notions require empirical verification, because only by the systematic study of punishment will psychologists be able to know if, or under what circumstances, undesirable side effects occur following punishment. Any undesirable side effects are naturally determined, and therefore the laws governing them theoretically may be discovered. The reasoning is similar regarding the transient effects of punishment. Only by empirical investigations will the principles of punishment be established and only then will evaluative judgments be authoritative.

Advocating the study of punishment, of course, is totally different from advocating its improper use in our society. There are certainly enough unresolved questions to warrant caution. However, enough work has been done to clarify many of the underlying principles of punishment, and psychologists are beginning to use it successfully in some situations (see Chapter 10 for a discussion of this work). In a recent article, Johnston (1972) summarized this point by saying, ". . . unconditioned and conditioned punishing stimuli as consequences to behavior delivered by our social and physical environment are as much a natural part of our lives as are positively reinforcing consequences. This being the case, behavioral science should undertake to understand and to control the results of their use [p. 1051]."

Thorndike's Truncated Law of Effect

In Thorndike's early writings, the Law of Effect had two aspects, one dealing with the strengthening of a response by "satisfiers," and the other with weakening a response by "annoyers." Thus, the Law of Effect dealt with both reward and punishment. Furthermore, the rewarding and punishing effects of stimuli were equal but in the opposite direction. A response that was "stamped in" by a positive reinforcer could be likewise "stamped out" by a noxious event. The important point is that both events changed response strength to a considerable degree but in the opposite direction.

In 1932, however, Thorndike made a radical change in his theory. On the basis of several experiments, he concluded that the Law of Effect did not apply to the effects of punishing operations. According to Thorndike, punishment was not effective and therefore was not the analogous, but opposite, process to reward. He retained only the positive aspect of the Law of Effect, and stated that the predominant effect of a punisher was to divert or shift the subject to another response rather than to actually weaken the initial punished response.

Among the many experiments cited by Thorndike to support his contention was one on verbal learning (Experiment 71). Subjects were asked to identify the correct meaning of Spanish words by choosing one out of five alternative response words on a list. If the subject chose the correct answer, he was rewarded by the experimenter saying "right." If the subject was incorrect, the experimenter said "wrong." Rewarded responses tended to increase in frequency; as such, the subjects chose the correct items more frequently than was expected by chance. In contrast, punished items were not weakened or suppressed; they were not avoided or ignored on later trials but rather continued to be chosen with the same relative, chance frequency as the remaining neutral words. Thorndike therefore concluded that punishers, compared to positive reinforcers, were ineffective since they failed to weaken responses. Rather, they worked indirectly by making the subject choose another response.

Similar conclusions were subsequently reached by other investigators. For example, Skinner (1938) punished rats for pressing a lever by having the lever quickly snap upward thereby slapping the rats' paws. He found that the effect of this punisher was only temporary. In a widely cited monograph, Estes (1944) reported a variety of experiments on punishment. Using a CER technique, he studied the effect of contingent versus noncontingent shock on the suppression of lever-pressing. Both types of presentation produced a temporary reduction in response rate, but the effect was generally weak and transient. Estes concluded that contingent shock punishment created a generalized disruption (like noncontingent shock) but, because the effects lasted only a short time, it did not weaken response strength.

Recent Evidence on Punishment Contingencies

A great deal of recent evidence, however, has refuted these earlier findings, principally by showing that a punisher is, in fact, a highly effective modifier of behavior when made contingent upon a response (see Church 1963, 1969, for reviews of this literature). For example, Hunt and Brady (1955) trained rats to press a lever for food on a variable-interval schedule. During the presentation of an S_Δ, one group received shock contingent upon lever-pressing. A second group received shock at the S_Δ offset regardless of whether the subjects had pressed the lever or not. The authors found almost complete suppression during the S_Δ for the contingent group but not for the noncontingent group.

Similarly, Azrin (1956) trained pigeons to peck a disc on a variable-interval schedule. During the S_Δ presentation (a change in the disc color), several shock schedules prevailed. On one schedule, shock was delivered at fixed intervals but it was not contingent on the pecking response, while for a second schedule, shock was delivered at the same fixed intervals but it was contingent on the response. Azrin found that contingent shocks were more effective in producing response suppression. Furthermore, the effects of the schedule on the pattern of responding were the same as those for reward conditioning but in the opposite direction. For example, a scalloping effect was found for the fixed-interval schedule where the response rate was high following shock and gradually decreased until the next shock.

The basic conclusion from these and other studies is that while noncontingent shock may disrupt ongoing behavior somewhat, its effect is relatively weak. Contingent punishment, on the other hand, is capable of totally suppressing behavior.

A third illustration of this point is given in Figure 6-1. Camp, Raymond, and Church (1967) first trained rats to press a lever for food during an S_d; a response produced food and terminated the S_d. Following the response they administered to separate groups either no shock, contingent shock, or noncontingent shock, given randomly between S_d presentations. Figure 6-1, showing the mean percentage of the S_d presentations during which the subjects responded, clearly indicates the suppressive effects of a contingent punisher in comparison to those of a noncontingent punisher.

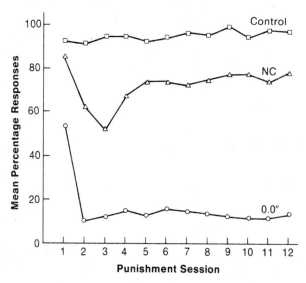

Figure 6-1. Mean percentage of responses as a function of no punishment (control), noncontingent punishment (group NC), or contingent punishment (group 0.0″).

Theories of Punishment

In the preceding discussion, three theories of punishment were implicitly presented. The first was Thorndike's original theory stating that the response itself was weakened if punished. The second theory,

Thorndike's later formulation, discounted punishment as effective, at least in the sense that positive reinforcers are effective. Rather, punishment acted indirectly by making the subject engage in other behaviors. According to this view, the response appeared to be weakened but, in fact, the subject was performing other responses.

Related to Thorndike's notion was the third theory, offered by Estes. In his studies, contingent and noncontingent shocks were essentially of equal effectiveness in suppressing behavior. This finding led Estes to propose that the punishing shock, in general, had produced a conditioned emotional response in the subjects—a generalized disturbance which resulted in withdrawal. Furthermore, this emotional state was aroused by the stimuli in the situation, including those discriminative cues that were explicitly presented by the experimenter, i.e., the S_Δ. Estes claimed that the fear state, elicited by the stimuli, disrupted the behavior and therefore the stimuli-shock relationship, as opposed to the response-shock relationship, was of primary importance.

The findings previously discussed on the effect of contingent versus noncontingent punishment indicated that Estes' theory had to be revised. It was clear that response-produced stimuli (internal stimuli that are generated when a response is made), as opposed to other external cues, are important in the analysis. Because immediate, contingent shock is more effective in suppressing behavior than a noncontingent punisher, the conditioning to internal stimuli is clearly more crucial than Estes believed it to be. To summarize, all stimuli, but especially response-produced stimuli, which are followed by shock punishment acquire the capacity to suppress behavior. Response-contingent punishment, however, is more specifically related to the behavior itself (i.e., the response-produced stimuli), and therefore produces greater suppression.

The fact that subjects engage in another activity after being punished is obvious. The reason they do so is not fully known, but several possible explanations have been offered. Estes' position, and its modification to include the greater importance of response-produced stimuli, was that the aversive stimuli elicited a state of fear which disrupted behavior. The position does not specify the exact nature or cause for the disruption other than to say that fear is disruptive. Several theories have been stated, however, which do suggest further principles.

The first was the competing-response hypothesis proposed by Guthrie (1934). His claim was quite simple: Aversive stimuli elicit unconditioned skeletal responses which are physically incompatible with the ongoing behavior. Shock naturally elicits freezing or locomotion responses that cannot be performed without disrupting other forms of behavior, e.g., lever-pressing. According to Guthrie, the concept of fear was unnecessary. Rather, the skeletal responses *per se*, which occur in response to aversive stimuli account for the disruption.

Guthrie's theory has merit in the sense that aversive stimuli do elicit unconditioned responses that would compete with the learned behavior. However, by limiting the range of unconditioned responses to skeletal ones, the theory has some difficulty dealing with the principle of acquired fear and, therefore, the possibility that subjects might be reinforced for engaging in other behaviors that reduced fear. Conditioned fear, as a consequence of punishing operations, seems plausible; if so, there certainly must be opportunities for fear reduction, by not engaging in the punished behavior, which in turn would be reinforced.

This latter notion is at the center of Mowrer's (1947, 1960) two-factor theory as it applies to punishment. Fear is first conditioned, according to classical conditioning principles, to the response-produced stimuli, and is subsequently elicited by the response-produced stimuli when the subject engages in that response. By not responding, fear is reduced, or alternatively, not responding is negatively reinforced. Therefore, response suppression is observed following punishment not simply because other responses are elicited, but rather because other responses—in fact, nonresponding—are explicitly reinforced by fear reduction.

A final theory to be considered is one suggested by Dinsmoor (1954, 1955), whose position represents a slight modification of the avoidance or two-factor position. Dinsmoor claimed that two factors were required to explain the effects of punishment, although he did not advocate labeling one as fear. Rather, according to his view, the discriminative stimuli which normally evoke the response, are paired with the noxious stimulus and consequently gain an aversive property in their own right. Any response which allows the subject to terminate or avoid these aversive stimuli is negatively reinforced. The difference between this explanation and Mowrer's theory is a subtle one. In both cases, the suppressive effects of punishment are due to the conditioning of avoidance reactions. However, Dinsmoor denied that the concept of fear or anxiety (or more specifically fear reduction) is needed to account for the effect.

The two-factor explanation, based on the negative reinforcement of fear-elicited withdrawal behavior, is extremely difficult to substantiate. One reason, according to Rachlin and Herrnstein (1969), is that nonresponding is not equivalent to responding; it is not a specified behavior and thus its increment due to negative reinforcement is a matter of interpretation. Unlike avoidance responding which is observable, an increase in the strength of a nonresponse during punishment is not directly observable. More important, the theory cannot account for the slight suppression of responding or, more correctly, the slight increment in a nonresponse, due to noncontingent punishment.

Rachlin and Herrnstein (1969) give evidence that punishers (empirically) have the same effects on responding as positive reinforcers, but in the opposite direction. From a descriptive point of view, then, these findings are best described by Thorndike's original formulation, although his proposed mechanism (e.g., response weakening) is not appropriate. The reason for rejecting Thorndike's mechanism is, in part, a logical one. If the response was, in fact, being weakened, suppression in the presence of the S_Δ but not the S_d (e.g., Azrin's [1956] study) would not be expected. Rather, a general decrement in performance should occur during both the S_Δ and the S_d. If any mechanism or underlying cause is appropriate for explaining the suppressive effects of punishers, it would seem more likely that the species-specific defense behavior in response to fear stimuli would provide a more useful focus than the negative reinforcement of avoidance behavior (nonresponding).

In conclusion then, although the symmetrical Law of Effect has been supported empirically, the reason for the finding remains unclear. Possibly, it is related to the natural, species-specific reactions to aversive stimuli, although it is also possible that nonresponding is negatively reinforced.

Suppression of Appetitive Responding

As indicated earlier in this chapter, punishment normally suppresses appetitive behavior. The following factors—punisher intensity, delay of punishment, duration of punishment, and prior stimulation—contribute to this effect.

Punisher Intensity

Suppression of responding is generally related to the intensity of the punishing stimulus, as demonstrated in numerous studies (e.g., Azrin & Holz, 1961; Camp, Raymond, & Church, 1967; Karsh, 1962, 1964). In the Camp et al. study, rats pressed a lever to obtain food on a VI schedule. During 10 punishment sessions, subjects were given a 2-second shock following a response at minute intervals. The shock intensity ranged from 0 to 2 mA for separate groups.

As shown in Figure 6-2 (refer to 0-seconds groups), the suppression of responding increased as a function of shock intensity. Suppression of responding was measured by the number of responses during

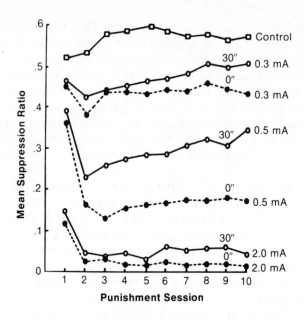

Figure 6–2. Mean suppression ratio as a function of no punishment (control), different intensities of shock (groups .3, .5, and 2.0 mA), and delay of punishment (groups 0 and 30 seconds).

the punishment session divided by the total number of responses during the punishment and prior nonpunishment sessions. Whereas the .3-mA group was only slightly more suppressed than the controls, the 2.0-mA group showed almost complete suppression. In summary then, shock intensity appears to be an important variable affecting the degree of response suppression.

Delay of Punishment

Delayed punishment attenuates suppression, as illustrated in Figure 6-2 (compare the 0- and 30-second delay groups). At each shock intensity, the groups receiving a 30-second delay of punishment showed less response suppression than the corresponding 0-second delay groups. Other investigators have also shown that delay attenuates the suppressive effects of punishment (e.g., Baron, 1965).

Duration of Punishment

The parameter of punishment duration has received relatively less attention, but some experiments indicate that longer durations of shock produce greater suppression. Boroczi, Storms, and Broen (1964) and Storms, Boroczi, and Broen (1963) showed that while a lever-press response was totally suppressed more quickly with longer punishment durations, the recovery of the response after punishment was discontinued was not related to shock duration. This finding is not consistent with effects found for intensity of shock (Boroczi, Storms, & Broen, 1964), i.e., recovery from punishment is inversely related to intensity. The reason for this dissimilarity is not clear, although it is probably related to the fact that the salient event in fear conditioning is the CS-US onset relationship (Mowrer & Solomon, 1954) rather than the duration of shock (i.e., US offset).

Prior Stimulation

The suppressive effects of punishment are not due exclusively to the momentary intensity of the punishing stimulus. Rather, previous experience with punishment influences later effects, as shown by Miller (1960). Rats were allowed to traverse a straight alley to procure food in the goal box for 64 trials, by which time they were performing at a high, stable rate. Two separate groups were then designated. On trials 65 to 150, one group continued to receive food for the running response. The second group was given a .1-second shock, as well as food, in the goal box, according to a set pattern: Namely, shock intensity was initially quite low (125 volts through a 250-K ohm limiting resistor) but was increased gradually by 15 percent after every 5 trials. By the end of their training (trial 145-150), this group was receiving a 335-volt shock. Both groups were then tested for 20 additional trials on which they were given a strong (335 volts) punishing shock at the goal. This was the first experience with punishment for the "sudden" group, but a continuation of the strong shock for the "gradual" group.

Miller measured speed of running, which, as shown in Figure 6-3 (on next page), was gradually reduced during trials 65-150 for the "gradual" group, while it was not affected in the "sudden" (unshocked) group. The interesting results pertain to the subsequent effects of strong punishment. Response speed was predictably suppressed in the "sudden" group. In contrast, the "gradual" group had acquired the capacity to resist the pain and/or fear associated with strong shock and therefore those subjects were relatively unaffected; they continued to respond despite the fact that they were receiving the same strong punisher, because of their previous experience with gradually increasing punishment. In a second experiment, Miller (1960) showed that gradually increasing shock, given in a separate apparatus, did not increase resistance to stress; rather the effect was limited to punishment presented in the goal box.

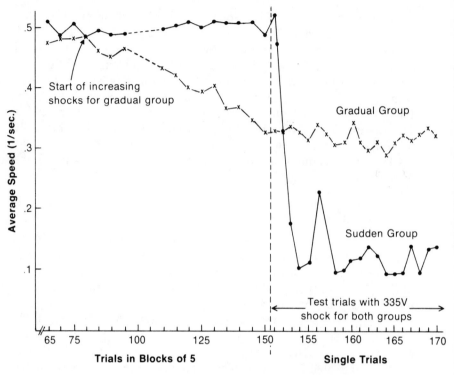

Figure 6–3. Mean running speed during training as a function of the gradual or sudden presentation of shock.

Miller hypothesized that the subjects had learned to tolerate stress through counterconditioning—running despite the gradually increasing shock was reinforced with food presentation. Therefore the response was sufficiently strong to be maintained later when the shock was strong. These results were confirmed and extended by Feirstein and Miller (1963) who showed that the locus of shock in the alley is important.

Facilitation of Appetitive Behavior

Although the basic effect of punishment is response suppression, there are numerous studies which indicate that paradoxically punishment may facilitate some responses. The most widely investigated effect is facilitation of T-maze learning when shock is given for the correct choice (see Fowler & Wischner, 1969, for a review).

Muenzinger (1934) was the first to note this phenomenon. Four groups of animals were trained to make a correct turn in a T-maze in order to obtain food. One group was given a mild shock in the correct arm of the T-maze, a second group was given a mild shock in the incorrect arm, a third was given food but no shock, and a fourth, the control group, was given neither food nor shock. The basic effect which Muenzinger found was that although the shock-wrong (S-W) group was slightly superior to the shock-right (S-R) group, the difference was not statistically significant. Both groups, however, were significantly superior to the group that received food only. The results indicate, therefore, that shock had a facilitating effect upon T-maze learning rather than a suppressing effect, and that the facilitation was independent of whether the subject was punished for the correct or incorrect response.

Muenzinger theorized that the shock had an alerting function; it slowed the subjects down so that they paid closer attention to their surrounding cues which improved learning. Several subsequent experiments tended to support this hypothesis. For example, Muenzinger and Wood (1935) showed that shock given for both responses was as effective in facilitating choice learning as a shock given for either the correct or incorrect response alone. Furthermore, these groups were superior to both a no-shock group and a group given shock prior to the choice point in the T-maze.

In two additional studies, the hypothesis of a general alerting function of shock was supported. Requiring the subjects to jump a gap in the floor of the maze after the choice point (Muenzinger & Newcomb, 1936), or detaining the subjects (Muenzinger & Fletcher, 1937), both facilitated learning.

A number of subsequent studies, however, criticized Muenzinger's conclusions as well as his method. Muenzinger, Bernstone, and Richards (1938) showed that if the shock was turned off for the S-W group following an incorrect response, performance was superior to that of the S-R group; although, again, both groups were superior to the no-shock controls. In Muenzinger's original study, shock was given continuously once a response was made, and since the S-W group could retrace and go to the correct side following a wrong turn, those subjects were obtaining twice the amount of shock that the S-R group was receiving. Therefore, this later study by Muenzinger et al. (1938) eliminated this problem and demonstrated that the shock performed two functions: First, there was a specific suppressive effect of shock which facilitated the performance of the S-W group relative to that of the S-R group; second, there was the general alerting function of shock which accounted for the superiority of the S-R group relative to the no-shock group.

In all of the studies by Muenzinger, a correction procedure was used; the subjects were allowed to correct a wrong response by retracing and going to the correct side. This procedure presented a major methodological problem: The S-W subjects often had a combination of rewards (e.g., escape from shock, as well as food) and consequently, on some trials, learned not only to approach the correct side but also to avoid the incorrect side. In contrast, the S-R subjects did not receive shock for an incorrect response, and therefore developed less avoidance for that side. Regardless, Muenzinger's procedure made it difficult to separate the learning-to-avoid the incorrect response from the learning-to-approach the correct response, because subjects were allowed to make both responses on the same trial.

Wischner (1947) corrected this fault by attempting to replicate the conditions used by Muenzinger, but using a noncorrection procedure. Once the subject had made the choice, the trial was terminated and the subject was removed from the apparatus until the next trial began. Wischner found superior learning for the S-W group relative to the S-R and no-shock groups, but noted that these latter groups did not differ from one another. In fact, the S-R group initially made many errors and only toward the end of training was learning evident. Wischner concluded that shock may slow the subjects' running, but its specific effect, suppression, was predominant. Wischner's experiment therefore strongly suggested that facilitation does not occur, at least when a noncorrection procedure is used.

A great number of studies have been done recently to specify more accurately both the facilitative (general alerting) and suppressive (as in Wischner's study) effects of shock on simple T-maze learning. Many of these have been parametric investigations of such variables as shock intensity (Wischner, Fowler, & Krushnik, 1963), shock duration (Wischner & Fowler, 1964), and the difficulty of the T-maze choice (Fowler & Wischner, 1965). In each case, the performance of the S-W group was superior to the S-R group, and this difference tended to be magnified with more intense or longer shocks. These findings suggest that shock produces fear, and that the conditioned aversive stimuli consequently increase errors for the S-R group and reduce errors for the S-W group.

However, as demonstrated by Fowler and Wischner (1965), shock has a second facilitating function. Rather than generally alerting the subject, as believed by Muenzinger, shock also is an additional cue which increases the discriminability of the alternative choices. Its facilitating effect is most apparent for more difficult discriminations where the other available cues are less salient. Under those conditions, the shock cue becomes relatively more important in allowing the subject to discriminate the incorrect from the correct response.

This effect was clearly shown in a second experiment on task difficulty (Fowler, Spelt, and Wischner, 1967). Hungry rats were trained to choose the correct arm of a T-maze to obtain food. The correct choice was easy when one goal box was fully illuminated while the other was darkened but the visual discrimination was more difficult when the darkened goal box was partially illuminated. Errors, noted for 360 trials, indicated that the S-R groups differed greatly from the no-shock groups when the discrimination was difficult. In contrast, differences between the groups were smaller when the discrimination was easy. Fowler et al. (1967) concluded that the cue function of shock was more important when the discrimination was difficult; therefore the S-R groups that received this helpful cue were superior to the no-shock groups that didn't receive the additional shock cue. With an easy discrimination problem, however, there was less need to utilize the shock cue, and thus the group differences were minimal.

To summarize, it seems clear that shock, given during discrimination learning, may have an adverse as well as facilitating effect. However, the reason for facilitation is that shock functions as a cue to increase the discriminability of the available choices rather than to generally alert the subjects.

Suppression of Aversively Motivated Responses

As is generally true with appetitive behavior, shock used as punishment suppresses both escape and avoidance responding.

Shock Intensity

One major variable that determines the degree of suppression is the shock intensity. Curiously, few experiments have investigated the effects of punishing avoidance acquisition. However, this procedure would be identical to avoidance studies (e.g., Kamin, 1954; Bolles, Stokes, & Younger, 1966) in which the subjects could not terminate or avoid the shock, since the continuation of shock for some time after the response was made could be considered as punishment. In those studies, it was clear that such a procedure did suppress behavior to a considerable degree.

The more typical design has been to punish an escape or avoidance response during extinction. Smith, Misanin, and Campbell (1966) first trained rats in a shuttle box to make up to eight avoidance responses. During extinction, they punished avoidance responses only (if the response was not made within the 5-second CS, the trial was terminated and no shock was presented). The authors varied shock intensity (0, 45, 73, 115, 185, or 300 volts through a 150-K ohm limiting resistor) and shock duration (.15 or 2.0 seconds). Eighty extinction trials were given.

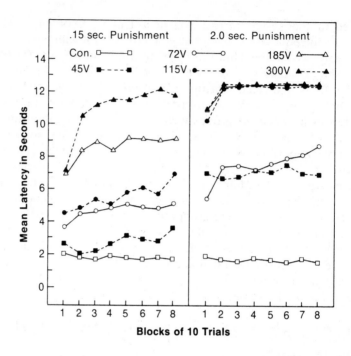

Figure 6–4. Mean latency during avoidance extinction as a function of shock intensity and duration.

As shown in Figure 6-4, latency generally increased over extinction as a function of both shock intensity and duration. The more intense or the longer the punishment, the greater was the suppression of avoidance behavior.

Punishment Duration

Several of the studies discussed earlier demonstrated that longer durations of punishment suppress aversively motivated behavior more than short durations. This was clearly shown in another study by Misanin, Campbell, and Smith (1966) in which rats were first trained to avoid a shock in a shuttle box. During 80 extinction trials, different groups were punished with either a .15-second or a 1.5-second shock (115 volts through a 150-K ohm limiting resistor) after an avoidance response. Punishment was delayed either 0, 2, 5, or 10 seconds for separate groups.

The results are shown in Figure 6-5. Median number of avoidance responses during extinction (i.e., responses made during the 12.5-second CS) was far less with the longer duration of shock punishment, except when the shocks were given immediately after the response, in which case both durations were of equal effectiveness in suppressing behavior. It is also clear that punishment delays retarded this suppression. In fact, for the shorter duration shock, the group that received a 10-second delay of punishment was indistinguishable from the control group (triangle) that received no shock. Misanin et al. (1966) suggested that the suppression reflected both a specific effect of contingent shock as well as a more general emotional disruption. Thus, these results are analogous to those found in studies of the punishment of appetitive responding.

Punishment Delay

The Misanin et al. (1966) study showed that suppression of responding decreased when the punishment was delayed and, therefore, that these effects of delay paralleled those found when punishment, or reinforcement, are delayed in appetitive conditioning. A wider range of punishment delays was investigated by Kamin (1959). In two studies which differed only slightly in method, he trained rats to avoid shock in a shuttle box to a criterion of eleven consecutive avoidance responses. During extinction, avoidance responses made within the 10-second CS were punished with a 1.1 mA shock. The shock, however, was delayed either 0, 10, 20, 30, or 40 seconds following the response.

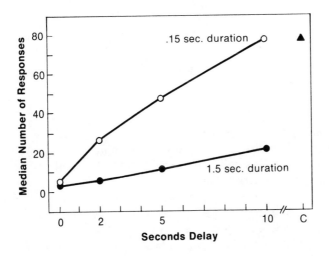

Figure 6–5. Median number of responses during avoidance extinction as a function of the duration and delay of punishment. The unshocked control group is indicated by the triangle at C.

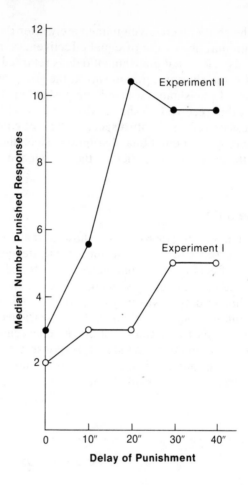

Figure 6–6. Median number of extinction responses as a function of punishment delay.

Kamin's results, shown in Figure 6-6, are especially dramatic in Experiment 2, where shock duration was more accurately controlled. When the shock was delivered immediately, the number of extinction responses was greatly suppressed; however, subjects continued to make more responses when the punishment was delayed for a longer time. The control group that was not punished made even more extinction responses than the groups for which punishment was delayed for 40 seconds, again indicating the presence of a general emotional factor which partially disrupts behavior. In summary, Kamin's study indicates that the delay-of-punishment gradient extends beyond 10 seconds (the longest delay used by Misanin et al., 1966) to 20 seconds, after which little additional effect is found with further increases in delay.

Facilitation of Aversively Motivated Responses

Punishment of escape or avoidance extinction may also produce facilitation of the response. This finding is paradoxical because punishment, as a suppressor of behavior, should hasten extinction rather than retard it. Nevertheless, under some circumstances, punishment may prolong responding during avoidance or escape extinction (see Brown, 1969, for a review). This phenomenon has been called self-punitive or vicious-circle behavior because continued responding during extinction leads to further punishment.

Self-punitive Behavior

An example of self-punitive behavior in escape extinction is provided by Brown, Martin, and Morrow (1964, Experiment 2). Rats were given 20 shock escape trials in a straight six-foot alley. Following, 60 extinction trials were administered, during which no shock was applied to the start box. Three separate groups were identified during extinction. The control group received no shock punishment, a second group (short shock) received shock punishment for running down the alley only when in the last two feet of the apparatus, while a third group (long shock) received punishment throughout the entire six feet of the alley.

The results for the entire six days of extinction testing are shown in Figure 6-7 (on next page). The slowest group in each of the two-foot segments of the alley was the unpunished group. Both punished groups continued to run faster throughout extinction with the long-shock group being faster than the short-shock group. The major point of the study is that shock punishment, given during the extinction of escape behavior, may enhance the response rather than suppress it. Not only were the punished groups faster than the no-shock group, but speed actually increased for the short-shock group as it got closer to the locus of punishment (the third two-foot segment).

In a study dealing with avoidance extinction, separate groups of rats were first trained to avoid shock in a seven-foot alley, then were given either no punishment during extinction, a brief punishing shock at the start of the alley, or a shock just prior to the goal box (Campbell, Smith, & Misanin, 1966). Thus, one major difference between this experiment and the one by Brown et al. (1964) was the shock duration.

The results of Campbell et al. (1966) were complex in that they showed both suppression and facilitation. The group punished outside the start box continued to avoid at a much higher rate than the no-shock control group. However, the goal-punished group extinguished more quickly than the control group. Campbell et al. concluded that facilitation is not always obtained; rather, shock early in the response sequence may produce self-punitive behavior but punishment later in the sequence may suppress behavior and thus hasten extinction. Another experiment, however, did find self-punitive behavior for both the short- and long-shock conditions (Martin & Melvin, 1964).

Theory

The most encompassing theory of self-punitive behavior was proposed by Mowrer (1947), who reasoned that fear was conditioned during acquisition and that it was, in turn, maintained (or augmented) by shock punishment during extinction. Whereas fear in the control group gradually extinguishes, fear is strengthened in the punished experimental group, and, therefore, motivation for responding is maintained.

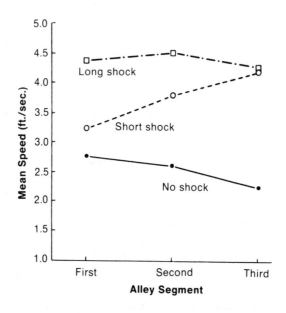

Figure 6–7. Mean running speed in each two-foot alley segment as a function of the shock placement and duration.

Mowrer's fear hypothesis accounts for much of the data, but the analysis is complicated by the fact that shock punishment, in addition to restoring fear motivation, also has other effects on the behavior. During shock punishment, the subjects continue to be reinforced by shock offset upon entering the goal box; such additional reinforcement is not available to the nonpunished group. It is therefore difficult to separate the effect of fear maintenance on self-punitive behavior from the effect of continued experience with shock offset (reinforcement), particularly in experiments like that of Brown et al. (1964) where rate of extinction was inversely related to the amount of shock.

This point was raised by Delude (1969). He claimed that the speed at which the subject left the start box was the appropriate measure of fear, since a strengthening of fear motivation due to punishment should increase the tendency to leave the start box. On the other hand, running speed in the alley where the punishment is administered would reflect the effects of the additional reinforcement due to shock offset. In his experiment, Delude found that self-punitive behavior was manifest using the alley-speed measure, but not using the speed with which subjects left the start box. He concluded that fear maintenance could not explain self-punitive behavior; instead, extinction was prolonged by the continued reinforcement of shock offset. A more recent study by Siegel, Melvin, and Wagner (1971), however, maximized the conditions for producing self-punitive behavior and took more accurate measurements of start-box speed. Under these conditions, start-box speed did show that extinction responding was prolonged by punishment.

In summary then, the enhancement of fear motivation does play a vital role in producing self-punitive behavior. The effect is noticeable on several measures, including trials to extinction as well as speed in the alley and start box. However, the effects of shock and shock offset during punishment extinction also influence self-punitive behavior.

There is no reason to exclude several other factors as possible determinants of the vicious-circle phenomenon. One of these variables is the degree of similarity between the acquisition and extinction sessions. According to this discrimination, or confusion hypothesis, extinction responding persists when the subjects are unable to discriminate acquisition from extinction. The confusion is greater for the punished groups, since shock is presented during both acquisition and extinction. In contrast, the unpunished control group extinguishes more quickly because one salient element—shock—is not present during extinction.

This theory is central to a number of studies, including Campbell, Smith, and Misanin's (1966). Only when the punishing shock was given near the start box (a situation resembling the acquisition conditions) was extinction prolonged. Several other authors, however, have discounted this hypothesis. For example, Melvin and Martin (1966) first trained rats to escape either a shock or a loud buzzer. During extinction, groups were given either no punishment, punishment with the same stimulus as presented during acquisition, or punishment with the other stimulus. The authors reasoned that if confusion led to self-punitive behavior, the groups that received a different punisher should extinguish more quickly than the nonpunished control groups.

The results of this experiment, shown in Figure 6-8, demonstrate that shock is more potent than a loud buzzer; i.e., the buzz-buzz group extinguished more quickly than the shock-shock group. More importantly, self-punitive behavior was demonstrated when punishment was presented during extinction regardless of which punishing stimulus was used. Shock and a loud buzzer increased extinction speed in different groups even when those groups had received the opposite stimulus during acquisition.

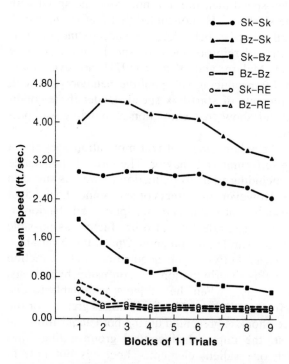

Figure 6–8. Mean running speed during extinction for the different groups that received either a buzzer (Bz) or shock (Sk) during escape acquisition, and a buzzer, shock, or no stimulus during punishment extinction.

A second approach was taken by Brown (1970). In his study, the visual or tactile cues in the center portion of the alley were changed during extinction. According to the confusion hypothesis, such a change should enhance discrimination, thereby reducing the confusion between acquisition and extinction conditions, and preclude self-punitive behavior. However, Brown demonstrated the vicious circle despite the change in environmental cues. These two studies strongly suggest that confusion between acquisition and extinction conditions is not the major determinant of self-punitive behavior. However, since it is impossible to independently verify the extent of confusion (as Brown, 1969, pointed out), such a factor may be one of several which contribute to the phenomenon.

In conclusion, the vicious-circle phenomenon represents a paradoxical effect of punishment and is related to a number of factors, including the locus and intensity of the punishing shock as well as the transition from acquisition to extinction. While a change in the nature of the punishing stimulus or environmental cues during extinction does not preclude the effect, vicious-circle behavior is enhanced when the transition is gradual. Finally, several factors undoubtedly account for the effect, including fear and confusion, although the former appears to account for more data than the latter.

Punishment and Neurosis

Relatively less is known about the conditions leading to neurosis. Part of the reason is that subhuman species, which are utilized in such experiments, convey only limited and nonverbal information. Psychologists have established conventions for inducing and measuring fear, but it is difficult to establish a general understanding of neurosis. The laws governing punishment and neurosis should become clearer with further research, but in the interim several areas have already been explored.

One neurotic behavior (in the sense that it is maladaptive and is based on fear) is the vicious-circle behavior discussed in the preceding section. By continuing to respond, subjects are exposed to additional shock; thus, the behavior resembles compulsive, or even masochistic, behavior. In fact, Mowrer's fear hypothesis was originally used to account for the fact that fear or guilt may perpetuate, rather than eliminate, fear-motivated behavior. Such an example is found in the enuretic behavior of children: Harsh punishment may prolong the behavior by increasing guilt or fear.

Several other phenomena have been investigated regarding punishment and neurosis. For example, in a study of response fixation following punishment (Maier, 1949), two groups of rats were trained to jump from a platform across an open space to one of two nearby goal boxes to obtain food. Each box had a swinging door which could be locked. If the rat jumped to the correct door it could get food, but a response to the incorrect door resulted in the rat falling to the net below and not receiving food. One group was taught to jump from the platform to a fixed position—the right door or the left door. A second group was given random reinforcement, sometimes on the right and sometimes on the left in an unpredictable order. Faced with such an insoluble problem, Maier found that the random-reinforcement group refused to jump. To force a response, the subjects were given an air blast or a shock. The result was that this group did jump but always in a very stereotyped way to one side. In the third phase of the study, Maier trained both groups to make a color-discrimination response: jump to a white versus black door. The first group had no difficulty learning this new task; the random-reinforcement group, however, continued to fixate to the one side throughout this phase regardless of the color of the door. Even when the jumping platform was adjacent to the two doors, or when the subjects could actually inspect the food behind the door, the fixation did not cease.

Maier hypothesized that the animals were in an extreme state of frustration, a state in which many learning principles don't apply. However, later formulations (Mowrer, 1950) claimed that the punishment had more simply increased the general drive level (fear) to an excessive point and that the subjects were therefore incapable of responding appropriately.

In a study supporting this latter argument, Farber (1948) trained rats to obtain food by choosing one side of a T-maze for 40 trials, followed by 60 additional trials during which they were shocked at the choice point. Farber then presented food in the other goal box and measured the number of trials required for the subjects to switch their response. He found that shock produced a fixation in the subjects; they continued to respond to the originally correct side rather than switch. Furthermore, fixation decreased if the subjects were fed at the choice point. Such a therapy-analog presumably had reduced fear and allowed the animals to learn the new response rather than continue to respond to the previous fear cues.

Fixation may result from punishment, and the effect is probably related to the extent of fear conditioned to the punishment cues as well as the conflict created by random reinforcement for the correct response. To the extent that such behavior is maladaptive, so-called neurotic behavior may be modified by procedures which change the strength or specificity of fear-related stimuli. Regardless, the notion that many neurotic behaviors such as fixation may be viewed in terms of learning principles, particularly those dealing with acquisition and extinction of fear, has received wide support (see Wolpe, 1952, 1953).

Finally, important contributions toward understanding punishment and neurosis were made by Masserman (1943) and Lichtenstein (1950), both of whom studied the effect of punishing consummatory (eating) behavior. In Lichtenstein's study, hungry dogs were shocked either when food was presented or during the actual eating response. Lichtenstein measured the number of trials required to inhibit the eating response as well as other, more general, behavioral changes both in the experimental situation and the home cage. Feeding was temporarily disrupted when the shock accompanied food delivery. However, when the dog was shocked while making the consummatory response itself, relatively permanent suppression of feeding was obtained. More importantly, a variety of neurotic behaviors accompanied this effect; some dogs became unusually passive and immobile in the apparatus while others became aggressive and hyperactive. In addition, many subjects developed tics and tremors, sudden changes in respiration, or retched in the presence of the food pellets. These dramatic symptoms led Solomon (1964) to conclude that punishment of consummatory behavior may uniquely lead to neurotic behavior, whereas punishment of instrumental responding may not have such effects.

Considerable work is required before a clear understanding of the relationship between punishment and neurosis will be achieved. At the present time, it may be concluded that neurosis is not an inevitable result of punishment. Neurosis may be linked to consummatory behavior or to a more general paradigm involving conflict and unpredictable stress (e.g., Weiss, 1968, 1971).

Summary

In 1932, Thorndike changed his earlier theory and suggested that punishment was relatively ineffective. Recent evidence, however, shows that a contingent punisher does suppress behavior to a considerable degree. According to one theory, punishers produce a general emotional state which disrupts ongoing behavior. Another theory holds that punishers elicit skeletal responses which are physically incompatible with the ongoing behavior. A third theory suggests that behavior is suppressed because nonresponding is actually reinforced and therefore increases in strength.

Punishment can both suppress and facilitate responding. Appetitive responding is suppressed to a greater degree by more intense, more immediate, or longer punishers. Prior punishment, if administered by gradually increasing its intensity, may result in a later resistance to intense punishment. In contrast, appetitive responding may be facilitated by punishment. If a correct T-maze response is punished, learning may improve; however, more recent studies indicate that such facilitation does not always occur. Learning may improve when the presence of a weak shock increases the discriminability of the two responses but, for easy discrimination tasks, the shock may suppress the response.

Punishment also suppresses aversively motivated behavior, with suppression increasing with more intense, longer, or more immediate punishers. Punished extinction may result in vicious-circle behavior because fear is enhanced by punishment, although the degree of confusion between acquisition and punished extinction may be a factor.

Although relatively little is known about punishment and neurosis, some research indicates that, during conflict or for a consummatory response, punishment may produce neurotic symptoms.

Extinction

Baum

Introduction

Most of the material discussed in previous chapters has concerned acquisition; however, extinction responding also provides an alternative measure of response strength. There has been a general belief, particularly as a result of Hull's (1943) influential book, that stronger habits should persist for a longer time in the absence of reward. As a measure of response strength, though, extinction responding often is not correlated with acquisition indices of habit. Nevertheless, a great deal of work has been done to determine the extent to which acquisition and extinction, as indices of response strength, are related. There is, moreover, a more fundamental reason for studying extinction: It provides an alternative focus for examining learning. The question of what happens to an acquired response when it is no longer followed by reward is a valid one in itself. By attempting to determine how responses are eliminated, it may become clearer as to how they are originally learned. Thus, the mechanisms of extinction are an important aspect of the study of learning.

Extinction is defined as the operation of withholding reinforcement following a response such that the subject is unable to obtain reward regardless of its behavior. The obvious consequence of this operation is a gradual decrease in the learned response and, conversely, an increase in alternative, competing responses.

As mentioned in Chapter 2, extinction does not completely eliminate a learned response. If a rest interval is provided following the extinction trials, the behavior spontaneously recovers, indicating that the response had become inhibited during extinction but was not eliminated. The nature of this inhibition and the factors involved are only partially understood at the present time.

Extinction of Instrumental Appetitive Responding

Any general theory of extinction would presumably have to account for the effects of a variety of factors. Several of these effects are considered below (see Smith & Bowles, 1971, for a bibliography of studies dealing with extinction).

Response Effort

One variable which influences extinction performance is the response effort. Mowrer and Jones (1943) were the first to demonstrate that greater response effort decreased resistance to extinction: Subjects performed more slowly in extinction with greater effort requirements.

This principle was confirmed more recently in a better controlled experiment by Capehart, Viney, and Hulicka (1958), in which rats were allowed to press a lever for food. Over the course of six days of acquisition training, all subjects experienced three different weights on the lever (5, 40, and 70 grams). During a subsequent extinction session, the subjects were divided into separate groups, and each received one of the three weights on the lever. It was clearly shown that responding (at least lever closures) ceased more quickly as a function of the effort. The 5-gram group continued to respond during extinction; of the other two groups, the 70-gram group extinguished more rapidly.

Amount of Reinforcement

Although responding is strengthened with larger rewards during acquisition, the opposite is generally true for extinction measures, i.e., extinction takes place more quickly in groups that had received larger rewards during acquisition. For example, Roberts (1969) trained rats to traverse a straight alley to obtain either 1, 2, 5, 10, or 25 pellets of food. During extinction, running speed was inversely related to previous reward magnitude in a monotonic fashion. These results have been obtained by numerous investigators (e.g., Hulse, 1958, and Wagner, 1961, although not by Hill and Spear, 1962).

The analysis of reward magnitude is complicated by the fact that it interacts with other factors in influencing extinction, a major one being the reinforcement schedule during acquisition. Larger, but consistent, rewards hasten extinction while the opposite is true for partial reinforcement schedules; extinction is prolonged following large reward on a partial schedule. This point is discussed more fully later in this chapter.

Amount of Prior Acquisition Training

The effect of training level in general is to increase resistance to extinction. For example, Williams (1938) examined extinction following either 5, 10, 30, or 90 reinforced lever-responses. Resistance to extinction increased as a function of the number of prior reinforcements. According to Hull (1943), this result indicated that acquisition and extinction measures were equivalent indices of response strength.

More recently, however, Ison (1962) obtained the opposite results while investigating extinction of an alley response. He found that a greater number of training trials decreased resistance to extinction. The reason for the discrepancy between Williams' and Ison's results is not entirely clear, although the probable cause is the difference in the reward magnitude. Magnitude and training level interact. For example, in a study by Ison and Cook (1964), separate groups of rats were trained to traverse a straight alley to obtain either 1 or 10 food pellets. One group in each condition was given 31 acquisition trials while the other was given 76 trials. All groups received 30 extinction trials. The authors found that the large-reward groups, in general, stopped running more quickly than the small-reward groups, which supported Ison's earlier experiment on reward magnitude. More significantly, greater training decreased resistance to extinction in the large-reward condition, while it increased resistance to extinction in the small-reward condition. This interaction explains the basis for the discrepancy, since the reward magnitude in the alley (.4 gram of food in Ison's study) was much greater than the one given in the lever box (.05 gram of food in Williams' study).

Delay of Reinforcement

The ways in which delay of reinforcement during acquisition affects later extinction responding are not thoroughly known; moreover, there are problems in interpreting the available data. Terminal acquisition performance is usually quite low for delay groups, suggesting that response strength is also low. Therefore, extinction data may simply reflect the factors which operated during acquisition rather than ones affecting extinction. Nevertheless, some principles have been formulated regarding the effect of reward delay on later extinction.

Some early studies showed that delay increased resistance to extinction. For example, Crum, Brown, and Bitterman (1951) rewarded one group immediately at the end of an alley but delayed the food presentation on half the trials for 30 seconds for the other group. As shown in Figure 7-1, the average latency for the immediate group (Group I) increased during extinction, while the delayed group (Group

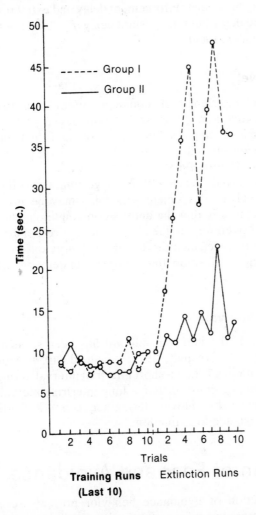

Figure 7-1. Running latency during acquisition and extinction as a function of immediate (group I) or delayed (group II) reward during acquisition.

II) continued to perform the response; a partial delay of reinforcement during acquisition had increased resistance to extinction.

This result has not been found in more recent investigations in which a constant delay of reinforcement was employed (e.g., Tombaugh & Tombaugh, 1969; Renner, 1965). Tombaugh (1966) found that resistance to extinction increased as a function of previous delay only if the subjects were removed from the apparatus immediately after each extinction response. In contrast, no effect of delay was observed when the subjects were detained during extinction for the same amount of time as on the previous acquisition trials. In summary, the relationship between reinforcement delay and extinction is unclear, although varied delay but not constant delay may increase resistance to extinction (Renner, 1964).

Drive Level

Drive level (e.g., length of food deprivation) has a strong effect on resistance to extinction. It is clear that extinction is prolonged not only when the drive level employed during extinction is high, but also when the previous acquisition drive was high (e.g., Barry, 1958; Campbell & Kraeling, 1954; Thieos, 1963). Conflicting results have been reported by Barry (1967) and Leach (1971) suggesting that other factors—intertrial interval, reinforcement schedule—may be involved. However, the consensus is that the deprivation conditions during acquisition do carry over and influence extinction responding, probably because subjects under high drive learn a slightly different and faster response during acquisition than subjects under low drive (Bolles, 1970).

Intertrial Interval

The effect of the intertrial interval has been studied by several investigators (e.g., Jernstedt, 1971; Sheffield, 1950; Stanley, 1952) whose general finding is that a short intertrial interval during extinction prolongs responding more so than a long intertrial interval. However, the intertrial interval employed during acquisition has little effect on subsequent resistance to extinction.

Extinction of Fear and Avoidance

The extinction of avoidance behavior presents several unique problems, illustrated in an important series of experiments by Solomon and his associates.

Anxiety Conservation

Solomon, Kamin, and Wynne (1953) first trained dogs, using a very high-shock intensity, to make an avoidance response of jumping over a hurdle into a safe compartment and then back again on the next trial. They found that once the response was established, it was extraordinarily difficult to extinguish. In fact, many subjects continued to respond during extinction for hundreds of trials even though no shock was ever presented. In order to facilitate extinction, the authors constructed a glass barrier over the hurdle which prevented the response and/or punished the subjects with shock for jumping. Both procedures were moderately successful in producing eventual cessation of responding, with the combination being more effective.

In a later paper, Solomon and Wynne (1954) developed the principle of anxiety conservation to account for this unusual persistence of avoidance responding. According to this principle, fear motivates avoidance behavior (Mowrer's two-factor theory). Since the subject continues to respond quickly during extinction prior to the time when the US would have occurred during acquisition, the Pavlovian fear never extinguishes. In other words, fear will extinguish only if fear occurs without being followed by the Pavlovian US (shock). By continuing to respond quickly, fear is not elicited or extinguished; therefore, fear motivation is maintained, and the avoidance response persists.

This principle poses a dilemma. The procedure used to extinguish fear (elicitation of the fear without the US presentation) encourages or motivates the continuation, rather than the extinction, of the avoidance response. What is clearly needed is a therapy, a procedure by which fear may be extinguished in the context of the avoidance behavior. Solomon et al. (1953) used punishment for some subjects, but, as discussed in Chapter 6, such a procedure may enhance fear and avoidance rather than suppress it. In fact, there is some evidence that vicious-circle behavior did occur in some subjects (Solomon et al., 1953).

The other therapeutic technique used by Solomon et al. (1953) was response prevention, or flooding. The authors attempted to extinguish fear by forcing the subject to remain in the presence of fear cues without the eventual US presentation. This form of reality testing is the basis for implosive therapy and has received a great deal of attention in recent years (see Baum, 1970, for a review).

Fear Extinction

It is understandable that many of the factors which influence fear extinction are those which influence fear acquisition. For example,

Baum (1969) found that higher levels of shock during acquisition decreased the efficacy of response prevention during extinction. He trained different groups of rats to avoid shock by jumping from an electrified grid onto a platform. After reaching a performance criterion, Baum confined the subjects to the grid area (response prevention or flooding) by retracting the safety platform from the box. Following either 0, 1, 3, or 5 minutes of the flooding procedure, the subjects were removed and then dropped again onto the grids, but no shock was presented. The persistence of the previous avoidance response was noted. Subjects that had received a higher shock intensity (either 1.3 or 2.0 mA) showed less extinction than subjects that had received a low intensity (.5 mA). Baum (1969) also showed that the duration of the flooding period was a major factor in determining the efficacy of the flooding procedure. Longer periods of response prevention facilitated later extinction of the avoidance.

In another study, Baum (1968) trained rats as described above, after which separate groups received either 0, 50, or 100 additional trials. He found that if the subjects did not perform their overtraining responses perfectly and thus received additional shocks, they took longer to extinguish than did the control-group subjects who received no overtraining trials. It is clear that the additional shocks had restored, or enhanced, their fear.

One final variable in facilitating extinction through response prevention was described by Lederhendler and Baum (1970). This experiment was like those cited above, except that during the 5-minute flooding period, the subject was physically disrupted by being nudged or forced by a small paddle to move around the apparatus. This procedure greatly enhanced the flooding and facilitated avoidance extinction. Lederhendler and Baum hypothesized that the procedure had interfered with fear expression, such as freezing, and had forced the subjects to develop relaxation responses.

Theory of Flooding

Lederhendler and Baum's (1970) finding questions the two-process theory of avoidance extinction which, as outlined by Solomon and Wynne (1954), claimed that Pavlovian fear must be extinguished during the flooding period before avoidance responding will extinguish. While much of the data are in agreement with this theory (e.g., shock intensity, duration of flooding period), it is not clear why mechanically forcing the subject to move around the apparatus would promote fear extinction.

A more serious criticism is that fear is not always extinguished during flooding. Coulter, Riccio, and Page (1969) found that avoidance responding extinguished more quickly following a flooding period, but subjects were still reluctant to approach a food cup that was located in the former shock box. In fact, subjects that received the flooding procedure demonstrated more residual fear than control subjects that did not receive flooding. This finding suggests that avoidance extinction is achieved with flooding by some means other than through fear reduction. The authors hypothesized that during flooding, subjects learn to freeze and crouch, which later competes with the active avoidance response and therefore facilitates extinction.

Baum (1969) has criticized this hypothesis on empirical grounds. During the 5-minute flooding period he noted different classes of behavior which the subject displayed—abortive avoidance responses, grooming, freezing, and general activity. As shown in Figure 7-2, freezing was relatively infrequent throughout the period, while general activity increased. Since the Coulter et al. hypothesis predicted that freezing would increase, Baum suggested that the subjects were learning relaxation responses. This term is somewhat confusing since its usual meaning is at variance with the Coulter et al. findings: There is residual fear even though the subjects may display general activity

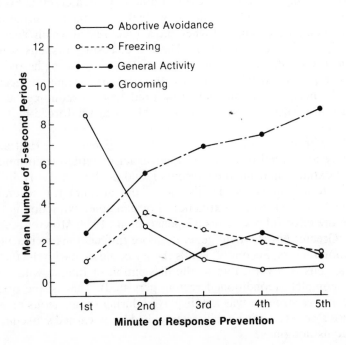

Figure 7–2. Mean number of 5-second periods during which a given activity was observed as a function of the time during the response prevention period.

during the flooding period. In summary, the nature and extent of the fear change during flooding is not clearly understood. Although response prevention, as a technique, does hasten avoidance extinction, there undoubtedly are a number of different determining factors.

Theories of Extinction

There are many explanations for extinction. Naturally, each succeeds on the basis of whether it accounts for the observed phenomena. One of the most elegant theories was developed by Hull (1943).

Inhibition Theory

Hull hypothesized that responding had two consequences: First, it increased the strength of the response, if reinforced; second, it produced an inhibitory tendency called reactive inhibition (I_r). The hypothetical construct I_r may be roughly equated with fatigue. During extinction then, the subject would become tired and would eventually stop responding; I_r would develop without any counteracting excitatory effect that is normally provided by the reinforcement. Furthermore, fatigue would be eliminated when the subject rested which, according to Hull, would be reinforcing. Therefore, resting responses which produced the fatigue reduction were increased in strength, and were termed conditioned inhibition ($_sI_r$). In summary, extinction occurred not only because the subject grew tired from responding but also because it learned to rest, a reaction which was reinforced by fatigue reduction.

As with other constructs in his theory of behavior, Hull unsuccessfully attempted to develop rather exact quantitative formulae to predict extinction, using the concepts I_r and $_sI_r$ (Jensen, 1961). In more general terms, however, Hull's theory did account for several important facts concerning extinction. For example, extinction is faster with more effortful responses (Capehart et al., 1958; Mowrer & Jones, 1943). Greater effort would generate more fatigue, thus increasing the inhibition factors. Secondly, Hull's theory could deal with the fact that spontaneous recovery, due to the dissipation of fatigue with rest, is rarely complete (conditioned resting persisted). That is, the original, full response strength was not displayed during spontaneous recovery because a counteracting habit (resting) had been learned subsequent to the acquisition phase.

There are serious problems, however, in attempting to account for other phenomena with Hull's theory. One such problem involves the massing of extinction trials, i.e., where trials are given quickly one after the other. Such a procedure increases resistance to extinction (Sheffield, 1950; Wilson, Weiss, & Amsel, 1955), although it appears to involve greater effort, or I_r, than a procedure in which the trials are distributed over a greater time period.

Hull's theory cannot account for the fact that simply placing the subject in an empty goal box will hasten subsequent extinction (Deese, 1951; Seward & Levy, 1949). This procedure, termed latent extinction, does not involve responding, and therefore there is no fatigue (see Moltz, 1957, for a review).

Last and most important is the criticism that Hull's theory cannot explain why partially reinforced subjects show greater resistance to extinction than continuously rewarded ones despite the fact that they are equal in terms of number of acquisition responses.

In conclusion, although Hull's theory has not been supported by recent research, it represented an insightful attempt to account for a complex issue with relatively few concepts. The theory has heuristic value and has effectively inspired numerous other studies which have formed the basis for alternative theories which are better understood in light of Hull's formulation.

Frustration Theory

The theory of extinction which is most widely supported is the frustration theory. A number of different investigators have advocated using the concept of frustration to account for extinction, although much of the impetus has stemmed from the work of Amsel and his associates (see Amsel, 1958, 1962, 1972).

The theory, simply stated, is that nonreward, when reward is expected, produces an internal state called frustration which is unpleasant and therefore motivating. Escape from or termination of the frustration is, in turn, reinforcing. Thus, frustration is a drive state which can energize behavior, and responses that lead to a reduction in the state are strengthened.

The energizing quality of frustration was demonstrated in a classic study by Amsel and Roussel (1952) in which hungry rats were trained to run down an alley to obtain food in a goal box and then continue running in a second alley to again obtain food in a second goal box. Running latency was measured in the second alley. Once running speed had stabilized, training was continued but frustration was introduced on half the trials; namely, on the frustration trials, the subjects were not given food in the first goal box but instead were detained for a short period of time. Amsel and Roussel reasoned that if nonreward was indeed frustrating and it led to an increase in motivation, behavior would be energized and running speed in the second alley would increase.

This prediction was confirmed (see Figure 7-3). On the frustration trials, running time was markedly less in the second alley than on the rewarded trials, clearly illustrating the motivational effect of frustration.

The response decrement observed during extinction stems from interfering or competing responses which are elicited by frustration. The capacity of such responses to compete with (or enhance) learned behavior was demonstrated by Adelman and Maatsch (1955). They trained rats to traverse an alley to obtain food. During extinction, the "jump" group was allowed to jump out of the goal box onto a platform. If subjects did not jump within 20 seconds, they were manually nudged by the experimenter. A second ("normal") group was simply confined in the goal box after an extinction response for the appropriate time period. The third ("recoil") group was allowed to retrace back into the alley upon finding the empty food cup in the goal box. After a subject did retrace, it was removed from the apparatus following the 20-second period. Therefore, the groups differed in the kind of response allowed following frustration. Adelman and Maatsch predicted that the "recoil" group would extinguish most quickly, since retracing into the alley was incompatible with the basic running response. On the other hand, the "jump" group would perform a response which was consistent with continued running and would show increased resistance to extinction.

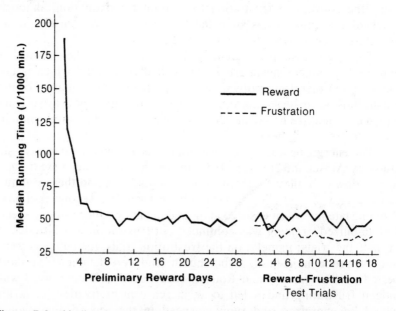

Figure 7–3. Median latency in the second alley as a function of reward or frustration in the first goal box.

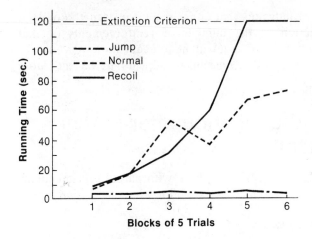

Figure 7–4. Running time during extinction as a function of allowing no response (normal), a jump (jump), or retracting response (recoil) following entry into the goal box.

The results, shown in Figure 7-4, clearly confirmed the predictions. The response latency for the "recoil" group increased to the criterion level within 25 extinction trials, whereas the fast performance of the "jump" group was unchanged after 30 trials.

In summary, the frustration theory can account for most of the major phenomena related to extinction and is particularly suited for explaining the partial reinforcement effect (the theory will be discussed more fully in that regard later in this chapter). The theory is an important contribution, since it successfully deals with the findings related to extinction on both an intuitive and empirical level. In doing so, it specifies such things as the cause and effect of interfering responses.

Partial Reinforcement Effect

One of the most perplexing phenomena in learning is the partial reinforcement effect. A vast number of experiments have been done and as many as nine major theories have been formulated to explain the effect (see Bitterman & Schoel, 1970; Lewis, 1960; and Robbins, 1971, for reviews). Any general theory of extinction, of course, must contend with the data pertaining to the partial reinforcement effect.

The basic effect is that partially reinforced subjects are more resistant to subsequent extinction than are continuously reinforced subjects. This finding indicates that either response strength does not increase as a function of the number of reinforcements or that resistance to extinction is not equivalent to acquisition measures of response strength. The latter possibility appears to be the more accurate statement.

Resistance to Extinction after Partial Reinforcement

A number of factors which combine with partial reinforcement schedules to influence resistance to extinction are considered below.

Percentage of Reward

The major factor determining the partial reinforcement effect is the percentage of rewarded acquisition trials. In general, resistance to extinction is inversely related to the proportion of rewarded trials. An example of this effect is illustrated in Weinstock's (1954) experiment in which hungry rats traversed an alley to obtain food on one trial per day. Seventy acquisition trials were given, and the percentage of rewarded trials was varied among groups. Running speed subsequently was measured during 20 extinction trials. Weinstock found that the continuous-reinforcement group extinguished almost immediately. The other groups, however, ran faster throughout extinction, particularly the 30-percent reinforcement group. The results confirm that resistance to extinction is an inverse function of percentage of reward during acquisition. This relationship has been demonstrated numerous times, although there are some studies in which an inverted U-shaped function was obtained (Grant & Shipper, 1952).

Reward Magnitude

As mentioned previously, a greater partial reinforcement effect is obtained when a large rather than a small reward is used during acquisition (Hulse, 1958; Leonard, 1969; Roberts, 1969). More specifically, large rewards administered on a continuous schedule decrease resistance to extinction while large rewards, administered on a partial schedule, increase resistance to extinction.

This phenomenon appears to be exclusively related to the effect of reward magnitude on extinction following continuous reinforcement. Roberts (1969) has shown, for example, that partially reinforced groups extinguish at about the same rate regardless of the previous reward size. In contrast, extinction for the continuously reinforced groups varied greatly according to reward size, with large rewards hastening extinction. In summary then, reward magnitude has little effect on extinction following partial reward. However, the partial reinforcement effect is larger with large reward because the continuous-reinforcement group extinguishes even more quickly under those conditions than after receiving small reward.

Recently this analysis has been further complicated by the fact that reward magnitude may also affect partially reinforced groups (McCain, 1970). Reward magnitude was varied in combination with training level (42 or 138 acquisition trials). All groups were rewarded on a partial schedule of 50 percent. McCain found greater resistance to extinction following acquisition with large reward after 42 trials but not after 138 trials. This finding strongly suggests that with relatively few trials, reward magnitude influences resistance to extinction in partially reinforced groups.

Training Level

In general, the effect of extended, partially reinforced training is an increase in resistance to extinction. This principle has been repeatedly demonstrated (Wagner, 1961; Wilson, 1964). For example, in Wilson's experiment, three groups (50-percent reinforcement schedule) received either 60, 180, or 240 trials in an alley. The median number of trials to the extinction criterion (three successive failures to enter the goal box within 20 seconds) for the groups was 15.0, 47.5, and 60.5 respectively.

Drive Level

Resistance to extinction after partial reinforcement increases with greater food deprivation (Mikulka & Pavlik, 1966) and decreases under conditions of satiety (Linton & Miller, 1951). Another finding is that partially reinforced groups respond more quickly during extinction than continuously reinforced groups following a change in drive from deprivation to satiation (Haas, Shessel, Willner, & Rescorla, 1970).

Shifts in Reinforcement Schedule

An interesting and important series of findings has been observed regarding extinction responding after shifting from one reward schedule to another during acquisition. Several investigators have demonstrated that a block of continuously rewarded trials presented just prior to extinction to a group initially rewarded on a partial schedule does not change the resistance to extinction relative to a group that received partial reinforcement throughout (Rashotte & Surridge, 1969; Thieos, 1962). Other investigators have found that either the continuous-partial group was superior to the partial-continuous group (Thieos & McGinnis, 1967), as well as the opposite effect (Leung & Jensen, 1968). Nevertheless, it is clear that shifted groups in general are more resistant to extinction than a group that received continuous reward throughout acquisition, and that even a few trials of partial reinforcement are sufficient to prolong extinction.

Patterning

Finally, an important variable determining extinction is the pattern of reinforced and nonreinforced trials during acquisition. An alternating pattern produces less resistance to extinction than a variable or random pattern when there are a moderate number of trials and when the number of trials and rewards are equated (Tyler, Wortz, & Bitterman, 1953). However, the opposite is true when a small number of trials are given. For example, Capaldi and Hart (1962) administered 18 trials of continuous, partial-random, or partial-alternating reinforcement to three separate groups and found that the alternating group was significantly more resistant to extinction than the other two groups. This apparent contradiction has been resolved in Capaldi's theory (see pp. 156-160).

Theories of the Partial Reinforcement Effect

As mentioned previously, numerous theories have been formulated to account for the partial reinforcement effect. Several of these are discussed in the following sections.

Expectancy and Discrimination Hypotheses

Humphreys (1939) was one of the first to investigate the partial reinforcement effect. He hypothesized that during acquisition subjects expect to receive reward; then, during extinction, a counter-expectancy develops before the subject stops responding. According to Humphreys, this counter-expectancy is slower to appear after partial reward because subjects have greater difficulty distinguishing acquisition from extinction. In contrast, the sudden change experienced by the continuous-reinforcement group allows the counter-expectancy to develop more readily.

The expectancy hypothesis never gained much favor, primarily because the concept of expectancy identified a psychological event without ever specifying the characteristics of that event, making the concept difficult to deal with experimentally.

A closely related, and more successful, theory was the discrimination hypothesis. Here, the cause for extinction was the subject's discrimination of acquisition from extinction. Discrimination was a somewhat more useful term in the sense that conditions may be varied and a differential reaction (discrimination) may be measured. The hypothesis explained the partial reinforcement effect by claiming that there was better discrimination of the extinction conditions following 100 percent reward than following partial reward. In other words, partial groups were confused during extinction and therefore persisted in their responding because they had experienced nonrewarded trials during acquisition, too. The theory also accurately predicted such findings as faster extinction for an alternating group compared to a random group (Tyler, Wortz, & Bitterman, 1953) where such a consistent alternating pattern is more highly discriminable from nonreward than is a random pattern.

An important study supporting the discrimination hypothesis was done by Bitterman, Fedderson, and Tyler (1953). Rats were run in an alley under a 50 or 100 percent reinforcement schedule. For one group, the goal box was the same on both the rewarded and nonrewarded trials, while for a second group, one goal box was used on rewarded trials and a second, discriminably different box on nonrewarded trials. During extinction, half of each group was given the rewarded box while the other half was presented with the nonrewarded box (or a novel box in the case of the 100 percent group that had experienced only a single box during acquisition).

The group that was most resistant to extinction was the one that experienced only one goal box on both rewarded and nonrewarded trials during acquisition and that same box during extinction. It was in that group that the similarity between acquisition and extinction was the greatest. The least resistant group was the continuously reinforced one that received the novel box during extinction since discrimination of extinction from acquisition was easiest for this group.

The discrimination hypothesis is intuitively plausible and has been supported by other experiments. However, it failed to account for other findings related to the partial reinforcement effect. For example, Marx (1958) gave reward to two groups of rats on a 100 percent schedule during acquisition. During extinction one group was given an empty food cup on each trial whereas the second group was given it on 50 percent of the trials. Marx reasoned that the latter group experienced the greater change from the acquisition conditions, and yet he found those subjects were more resistant to extinction.

Secondly, Thieos (1962) trained subjects for 70 trials on a 40 percent partial reinforcement schedule. Next, three groups received either 0, 25, or 70 additional reinforced trials, followed by 40 extinction trials. Thieos found that the partial-70 group was less resistant to extinction than the other two partial groups; i.e., the interpolated trials had reduced resistance to extinction somewhat. However, he also found that the partial-70 group was more resistant to extinction than a control group that had received 100 percent reinforcement throughout acquisition. Therefore, the 70 continuously reinforced trials were not enough to eliminate the partial reinforcement effect. Although this outcome may be a function of the degree of training, it does pose problems for the discrimination hypothesis in the sense that the transition from acquisition to extinction conditions should have been as discriminable for the partial-70 group as for the control group, and yet the groups differed on resistance to extinction.

Frustration Hypothesis

The frustration hypothesis has been very successful in accounting for extinction in general, and the partial reinforcement effect in particular. The theory states that a nonrewarded trial produces frustration and this response and the implicit stimuli it creates become conditioned to the surrounding cues. The internal, response-produced stimuli which are characteristic of frustration become part of the overall complex of stimuli that elicit the criterion response. During extinction, then, these cues continue to elicit the response. The partially reinforced subject has essentially been reinforced for responding in the presence of frustration cues and is thus conditioned to respond to nonreward during extinction. The continuously reinforced subject, on the other hand, has not previously experienced frustration, and is not conditioned to respond in the presence of the internal frustration cues which it experiences during extinction for the first time.

Although there are several sources of support for the frustration theory, it is less successful when trying to account for the partial reinforcement effect which is obtained following a very small number of training trials (Capaldi & Deutsch, 1967; McCain, 1966; Padilla, 1967). In these experiments, extinction differences were observed between continuous and partial groups after as few as 2 to 5 acquisition trials. It is unlikely that frustration would build up so quickly, and even more improbable that it would become conditioned so quickly to surrounding cues as Amsel and his associates have postulated. Although there have been some recent modifications of the frustration hypothesis to meet this challenge (Amsel, Hug, & Surridge, 1968), the evidence seems to favor adopting a sequential theory.

Aftereffects and Sequential Hypothesis

Capaldi's recent sequential hypothesis is an elaboration of a previous aftereffects theory postulated by Sheffield (1949). Sheffield reasoned that there are aftereffects, distinctive internal stimuli, which are produced on a nonrewarded trial and which persist to the start of the next trial. The stimulus complex or pattern, at the start of the response, comes to include these stimuli. On a rewarded trial (following a nonrewarded trial) the subject is reinforced for responding in the presence of these nonreward aftereffects. A continuously rewarded subject, however, never experiences nonreward aftereffects, and therefore is not conditioned to respond in their presence when they occur during extinction.

Sheffield tested her theory by varying the interval between trials. She predicted that the partial reinforcement effect would be obtained if the trials were spaced only 15 seconds apart (the nonreward aftereffects would easily be conditioned) but not if the trials were spaced 15 minutes apart since the nonreward aftereffects would dissipate within that time. Sheffield's predictions were confirmed (Sheffield, 1949).

A number of subsequent findings, however, have failed to support Sheffield's theory. First, several investigators (e.g., Lewis, 1956; Weinstock, 1954; Wilson, Weiss, & Amsel, 1955) have used widely spaced trials, i.e., 24 hours, and have still found a partial reinforcement effect. It is unlikely that the aftereffects of nonreward could persist for that long, particularly since Sheffield did not find the effect using even a 15-minute intertrial interval. Secondly, the aftereffects for an alternating schedule of reinforcement would be the same as for a random schedule (with equal rewards); yet, with extended training at least, the alternating pattern produces rapid extinction.

A more recent form of the aftereffects theory has been proposed by Capaldi (see Capaldi, 1966, 1967; and see Koteskey for a review). Capaldi hypothesized that nonreward aftereffects differ from reinforcement aftereffects and that they persist and become conditioned on the next rewarded trial. However, nonreward stimuli persist through memory and therefore do not decay rapidly. With this subtle but important change, the sequential hypothesis can explain almost all the findings regarding the partial reinforcement effect.

Capaldi's theory specifies three basic variables which influence the partial reinforcement effect. The first is the N-length, where N refers to a nonrewarded trial and R to a rewarded trial. Capaldi has found that resistance to extinction is greater with a longer series of nonrewarded trials. Presumably, the characteristic nonreinforced stimuli accumulate over such a series and are more intense on the subsequent rewarded trial.

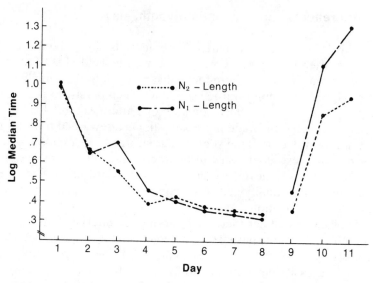

Figure 7–5. Median running time during acquisition and extinction as a function of N-length.

In an example of this principle (Capaldi, 1964, Experiment 2), two groups of rats received the same pattern of rewarded and nonrewarded trials for 8 days of training in an alley. For example, on the first day the sequence for both groups was RNNRRNNR. The only difference between groups was the location of an extra intertrial reward; in receiving an intertrial reward, the subject would simply be placed into the goal box and allowed to eat. Capaldi reasoned that such a procedure would restore the aftereffects of a rewarded trial without actually reinforcing the running response. Therefore, the sequence—number of N and R trials and the actual pattern—were held constant while varying the N-length. Group N_1 was given an intertrial reinforcement so that only one N-trial remained prior to the next R-trial (N-length was always one). Group N_2 was given an intertrial reward so that two N trials remained (N-length of two).

The results, shown in Figure 7-5, indicate that acquisition times were nearly identical for the two groups. However, group N_2 was much slower to extinguish than group N_1. This effect occurred on the basis of the difference in N-length alone.

The other two basic variables affecting resistance to extinction are the number of occurrences of an N-length and the number of different N-lengths. As mentioned above, aftereffects from N trials accumulate over a series of N-trials; the more times an N-length is reinforced, the stronger it becomes. In addition, the aftereffects from different N-lengths differ from one another and also tend to summate. Therefore, greater repetition of an N-length and a greater variety of N-lengths both may produce greater resistance to extinction.

These two principles were confirmed by Capaldi (1964, Experiment 3) when six groups of rats were run in an alley. Three groups were given patterns containing an N-length of three (e.g., RRNNNR), and the other three groups were given patterns with different N-lengths of either 1, 2, or 3 (e.g., NRNNRR or RRNNNR). In each of these conditions, one group was given a total of 24 trials, a second was given 60 trials, and a third 120 trials. Thus, the contribution of both factors—number of occurrences of a single N-length and number of different N-lengths—was assessed.

The results are illustrated in Figure 7-6. When only 24 trials were given, the N_3 group was more resistant to extinction. Capaldi explained this finding by stating that the strength of the single N-length for group N_3 was greater than the sum of the strengths of the three N-lengths for group $N_{1\ 2\ 3}$. In contrast, after 120 trials, group $N_{1\ 2\ 3}$ was more resistant to extinction than group N_3. Here, the summation of strengths eventually outweighed the strength of the single N-length for group N_3. The basis for these somewhat complicated changes is that the strength

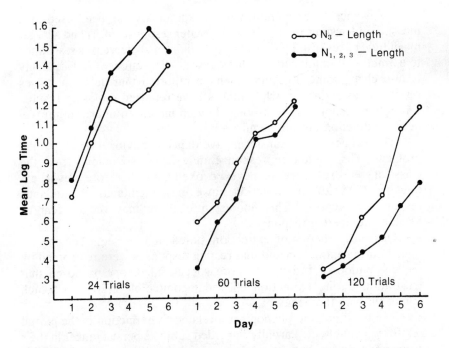

Figure 7–6. Mean running time during extinction as a function of number of acquisition trials and one or three N-lengths.

of any one N-length grows as a negatively accelerated, exponential function, i.e., there is a very large surge of strength at the beginning of training but further reinforcement adds less and less to this strength. Therefore, after 24 trials, the strength of the N-length for group N_3 had accelerated beyond the sum of the strengths of the three N-lengths for group $N_{1\ 2\ 3}$. However, once those N-lengths became strong, their sum was greater than the strength of the single N-length for group N_3.

It is clear that the three principles of N-length, number of occurrences of that N-length, and the variety of N-lengths, all interact. The experiments cited illustrate that the combination of these factors is important for determining resistance to extinction. In summary then, Capaldi's theory accounts for most of the phenomena regarding the partial reinforcement effect (e.g., patterns, percentage), and it can be easily applied to account for other findings such as the effects of reward magnitude (Leonard, 1969). The theory does have a few minor shortcomings, one being that it predicts acquisition speed to be lower than that usually found (see Koteskey, 1972).

Summary

Resistance to extinction, which traditionally has been used as a measure of learning, decreases with greater response effort and a larger magnitude of reward. Increased prior training also decreases resistance to extinction if combined with large reward. The effect of reward delay is less clear; some investigators have reported that delay increases resistance to extinction while others have reported opposing results. Extinction responding is prolonged by an increased intertrial interval during extinction but not acquisition.

The extinction of fear during avoidance behavior offers a unique problem: The avoidance response may occur so quickly that the classical fear response is never evoked and therefore never extinguished. Avoidance extinction, however, is facilitated by avoidance response prevention. This flooding procedure may reduce fear or heighten competing responses.

One early theory of extinction stated that subjects became fatigued during extinction and that resting responses were reinforced by the dissipation of fatigue. A more successful theory proposed that frustration during extinction elicited incompatible responses which competed with the behavior.

The most notable phenomenon related to extinction is the partial reinforcement effect: Partially rewarded subjects persist longer in their responding during extinction than continuously rewarded ones. This effect is an inverse function of percentage of reward during acquisition and is magnified by larger rewards and extended training. A series of partially reinforced acquisition trials, even though followed by continuously reinforced trials, increases resistance to extinction.

One of the first theories of the partial reinforcement effect was the expectancy theory, which stated that partially rewarded subjects learn to expect no reward during extinction more slowly than continuously reinforced subjects. A similar hypothesis, the discrimination theory, stated that partially reinforced subjects could not discriminate extinction from acquisition. Recent theories have been more successful. The frustration theory postulates that partially reinforced subjects, during acquisition, become conditioned to respond in the presence of nonrewarded stimuli which would not be true for the continuously reinforced subjects. Another interpretation, the sequential hypothesis, states that nonrewarded aftereffects are conditioned on subsequently rewarded trials and thus form part of the stimulus pattern which elicits the response during extinction. This theory predicts data largely on the basis of the number and type of transitions from nonrewarded to rewarded trials.

Secondary Reinforcement

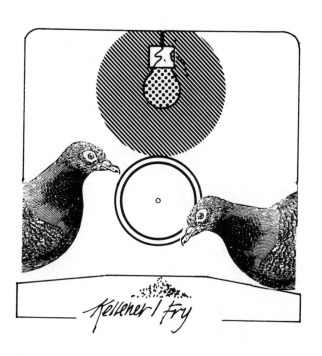

Kelleher / Fry

Introduction

Secondary reinforcement is one of the most important concepts in learning. A secondary reinforcer is an originally neutral stimulus which, through consistent pairing with primary reinforcement, acquires the properties of the primary reinforcer such that, on subsequent occasions, the secondary cue by itself will reinforce behavior. Some of the important questions regarding this phenomenon relate to how secondary reinforcement is developed, and how it operates on subsequent occasions to reinforce behavior.

The importance of secondary reinforcement, as a process, is clear, since complex human behavior is not always modified by primary reinforcement. Only on some occasions do humans perform instrumental responses for food as is typical in most learning experiments with hungry rats. Rather, secondary rewards, such as praise and money, are usually quite unrelated to, or at least removed from, reinforcers that are vital for biological well-being. However, as the above definition states, secondary reinforcers are, at least initially, related to primary rewards.

Secondary reinforcement is important to learning psychology for another reason: It has been the focus for significant research and theoretical debate specifically dealing with classical versus instrumental models of learning. Since the principle has been described in both classical and instrumental terms, this debate has helped to clarify the nature of secondary reinforcement, as well as classical and instrumental conditioning processes (see Hendry, 1969; Myers, 1958; and Wike, 1966, 1969, for reviews).

Experimental Designs for Demonstrating Secondary Reinforcement

Not surprisingly, the techniques for investigating secondary reinforcement are similar to those used to assess primary reinforcement. If a neutral cue does in fact act in the manner of a primary reinforcer, then it must influence behavior in a manner similar to the way a primary reinforcer influences behavior.

The three main techniques for determining if a cue is a secondary reinforcer include assessing its ability to (a) maintain responding, called chaining, usually in the absence of primary reward; (b) prolong extinction; or (c) serve as the reinforcer for the acquisition of a new response. The last method is in some ways superior to the other two, since it gives less equivocal evidence that the stimulus is, in fact, a reinforcer. These three methods are schematically represented in Figure 8-1. In each case, the stimulus (S_1) becomes a secondary reinforcer which is demonstrated subsequently during the test.

In the chaining technique, S_1 is classified as a reinforcer if it maintains the response in the presence of S_2. A response in the presence of S_2 is not initially reinforced. However, if the consequence of responding during S_2 is the presentation of S_1, and if such responding is, in fact, maintained, S_1 must be acting as a reinforcer.

The training and testing of secondary reinforcement by the second two methods is somewhat easier to visualize. During initial training in both cases, the stimulus is paired with primary reinforcement. For the extinction method, the stimulus is then presented alone during extinction and, if the stimulus is acting as a reinforcer, responding continues at a stronger level. This effect is relative to those control conditions in which either the stimulus is not presented during extinction (no reward whatsoever and therefore faster extinction), a novel stimulus is presented (one not previously paired with primary reward and therefore unable to act as a secondary reinforcer), or a stimulus is presented which previously followed the primary reward (thus producing little conditioning as discussed in Chapter 2 in connection to backward conditioning).

For the acquisition technique, the stimulus is first paired with primary reward, and then is subsequently presented following a second, novel response. If this second response increases in frequency or strength, the stimulus may be classified as a secondary reinforcer.

Technique	Training Procedure	Secondary-Reinforcement Test
Chaining or Response Maintenance	$S_2 \rightarrow R_1 \rightarrow$ no reward $S_1 \rightarrow R_1 \rightarrow$ reward	$S_2 \rightarrow R_1 \rightarrow S_1 \rightarrow R_1 \rightarrow$ reward
	Control conditions: $S_2 \rightarrow R_1 \rightarrow$ no reward $S_1 \rightarrow R_1 \rightarrow$ reward	$S_2 \rightarrow R_1 \rightarrow S_3 \rightarrow R_1 \rightarrow$ reward
Extinction	$R_1 \rightarrow S_1 \rightarrow$ reward	$R_1 \rightarrow S_1 \rightarrow$ no reward
	Control conditions: a) $R_1 \rightarrow S_1 \rightarrow$ reward b) $R_1 \rightarrow$ reward c) $R_1 \rightarrow$ reward $\rightarrow S_1$	$R_1 \rightarrow$ no reward $R_1 \rightarrow S_1 \rightarrow$ no reward $R_1 \rightarrow S_1 \rightarrow$ no reward
Acquisition of New Response	$R_1 \rightarrow S_1 \rightarrow$ reward	$R_2 \rightarrow S_1$
	Control conditions: $R_1 \rightarrow S_1 \rightarrow$ reward $R_1 \rightarrow S_2 \rightarrow$ no reward	$R_2 \rightarrow S_2$

Figure 8–1. Matrix schematically illustrating the relationships between responses (R), stimuli (S), and primary reinforcer (reward) during the establishment of a secondary reinforcer and the subsequent test of its effect by the chaining, extinction, or acquisition techniques.

Secondary Reinforcement in Appetitive Learning

Chaining Technique

The chaining technique for demonstrating secondary, or conditioned, reinforcement is widely used (see Kelleher & Gollub, 1962, and Kaufman & Baron, 1969, for reviews). As stated earlier, a response made in the presence of a previously unreinforced discriminative stimulus may be maintained by the presentation of a positive secondary reinforcer. An example is a study by Kelleher and Fry (1962), in which pigeons pecked a small plastic disc to obtain food. After a fixed interval of time during which the disc was illuminated by white light, a disc-peck changed the light from white to green. Again, after a fixed interval, a response turned on a red light. Finally, a response during the red light produced food reward. In summary then, the bird had three fixed-interval components, each designated by a separate color stimulus.

Kelleher and Fry (1962) found responding during all three components, although there were noticeable pauses in responding during the white stimulus. This effect is seen in the cumulative records shown in Figure 8-2. The completion of the three components, marked by the reinforcement blip, clearly indicates that pauses occurred prior to responding (at arrows in panels B and C). Furthermore, the slope of the curves indicating rate of responding is less at the start than later on in the sequence. Responding was slow during the white stimulus, but it accelerated during the green and red stimuli and was at its maximum just prior to reward.

In this experiment, the red stimulus was a first-order conditioned reinforcer because it maintained the responding during the green stimulus; otherwise responding would not have occurred without food reward. However, the green-stimulus presentation maintained responding during the white stimulus, although the effect was weaker. This exemplifies tertiary, rather than secondary, reinforcement: The green stimulus had become a conditioned reinforcer not by its direct association with food, but rather by its association with a secondary reinforcer. A tertiary reinforcer is a secondary reward once removed.

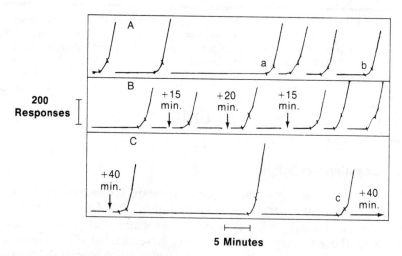

Figure 8–2. Cumulative record for one pigeon for sessions 80, 90, and 100 (panels A, B, and C respectively).

Humans are able to make use of conditioned reinforcers that are many times removed from primary reward, although lower animals are quite limited in this regard.

Extinction Techniques

There are numerous studies which illustrate secondary reward during extinction. For example, Miles (1956) trained rats to press a lever to obtain food. A response produced not only food, but also a light-flash and an audible click from the feeding mechanism. Different groups of subjects received either 0, 10, 20, 40, 80, or 160 reinforced responses during acquisition. Following, the subjects were extinguished to a criterion of 4 minutes without a lever-press. During extinction, half of each group was presented the secondary reinforcers—light and click—for each response while the other half was not. As shown in Figure 8-3, the number of responses during extinction was generally higher with the presentation of the secondary reinforcers.

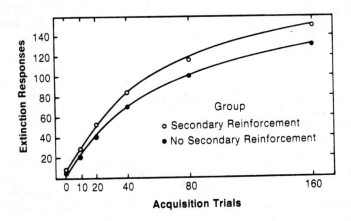

Figure 8-3. Number of extinction responses as a function of prior acquisition level and presentation of secondary reinforcement cues.

Acquisition Technique

In a classic study demonstrating secondary reinforcement using the acquisition technique, Saltzman (1949) gave hungry rats 25 acquisition trials in a straight alley. Group 1 was given continuous reinforcement in a distinctive goal box after traversing the alley; for some of the animals, the box was colored black, while for others it was white. Group 2 received reinforced trials alternated with nonreinforced trials.

The same color goal box was used for both the rewarded and nonrewarded trials. Group 3 was also given partial reinforcement, but their distinctive goal box was of one color for the reinforced trials and the other color for the nonrewarded trials.

The test for secondary reward took place in a separate apparatus —a T-maze shaped like a tuning fork. Here, the goal box previously used on the rewarded trials was at the end of one arm, while the nonrewarded goal box, or a novel box for group 1, was at the end of the other arm. Saltzman tested to see if the subjects could learn to make this new response (the correct turn in the maze) if given the secondary reinforcing goal box as opposed to primary reward. An important feature of these fifteen test trials was that the subjects could not see the goal box until after they had made their choice. Saltzman also employed a control group (group C) that received food reward for a correct response in the maze.

The results are shown in Figure 8-4. All the groups learned the new response and performed significantly above the chance level. Moreover, each group was as good as, if not better than, the control group that received food reward. These data illustrate that secondary cues may be potent reinforcers and that they are capable of meeting the most rigorous test of a reinforcer: They can serve as the basis for new response learning.

Secondary Reinforcement in Aversive Learning

It has been argued that cues associated with the termination of aversive stimuli become secondary reinforcers. This notion has important implications, since it indicates the generality of secondary reinforcement as a principle of learning applicable to both positive and negative reinforcement.

Chaining Technique

Dinsmoor and Clayton (1963) trained rats to press a lever during shock. That response produced a noise, and, 30 seconds later, a different response—pushing their nose against a small plastic disc—terminated both the noise and shock. The data indicated that the subjects did press the lever and that the reinforcer for this behavior was the noise. When the lever-press response no longer produced the noise, the frequency of pressing was virtually zero. Therefore, as in the previous example using appetitive reward, the stimulus occurring

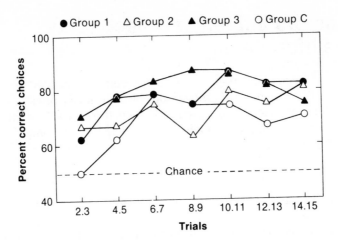

Figure 8–4. Percent correct choices during the acquisition of the maze response for the secondary-reinforcement groups as a function of trials.

before shock offset became a secondary reinforcer and maintained the behavior which preceded it.

Acquisition Technique

There are a number of recent studies employing the acquisition technique to demonstrate conditioned reinforcement based on shock termination. Kinsman and Bixenstine (1968) trained rats to escape a shock by running to the opposite side of the cage during a flashing light (CS_1). Immediately following the response, a second stimulus (buzzer) was given for .5 seconds (CS_2); the shock, CS_1, and CS_2 all terminated together. After 90 trials, a lever was inserted into the cage for the secondary reinforcement test. For a lever-press, different groups of subjects received either .5 seconds of CS_1, .5 seconds of CS_2, 5 seconds of CS_1 plus CS_2 during the last .5 seconds, or no stimulus. The results indicated that the presentation of CS_2 was reinforcing. The group receiving CS_2 only showed the highest frequency of lever-presses, followed by the group that received CS_1 plus CS_2. In summary, stimuli are capable of becoming secondary reinforcers by virtue of their association with shock offset. A study showing similar results was performed by Murray and Strandberg (1965). Recently, however, several investigators have questioned the existence of secondary reinforcement based upon shock termination (e.g., Beck, 1961; LoLordo, 1969; Siegel & Milby, 1969).

Factors Influencing Secondary Reinforcement

If secondary reinforcers, after being associated with primary reward, take on the characteristics of the primary reward and become capable of reinforcing responses, then the efficacy or strength of a secondary reinforcer should be influenced by the same factors that define the strength of a primary reward.

Acquisition and Drive Level

The Miles (1956) study cited previously suggested that the effect of greater acquisition was to increase the strength of secondary reward. The more times a cue is paired with primary reinforcement, the stronger is its power to function as a reinforcer. In that same experiment, Miles also investigated the effect of food-deprivation level on secondary reinforcement. Different groups of subjects were deprived of food for either 0, 2.5, 5, 10, 20, or 40 hours during extinction (all subjects had previously received 80 reinforcements for lever-pressing under 24 hours of deprivation). Resistance to extinction generally increased as a function of deprivation level. In addition, the groups that received the secondary cues—light and click—for their extinction responses were significantly more resistant to extinction than the comparable groups that did not receive the secondary cues.

Although the strength of the secondary reinforcer is greater when the subject is more deprived on the test, that strength is not related to deprivation level during acquisition (e.g., Brown, 1956; Hall, 1951; Wike & Farrow, 1962). Brown (1956) paired a light and buzzer with food. Groups of rats were under conditions of either high or low food deprivation at that time. On a later test, a lever was inserted into the cage and one half of the subjects received the secondary cues for a lever-press response, while the other half did not. A secondary reinforcement effect was shown, and although deprivation level during the test influenced responding, as found by Miles (1956), the strength of the secondary cues was not influenced by the prior deprivation level.

Schedule of Primary Reward

As indicated in Miles' experiment, secondary reinforcement is stronger with greater training. However, this effect depends, in part, on the schedule of reinforcement used during this acquisition training. A number of investigators have found that a partially reinforced cue becomes a stronger, more durable secondary reinforcer than a continuously reinforced cue (e.g., Armus & Garlich, 1961; D'Amato, Lachman, & Kivy, 1958; Klein, 1959). For example, Klein's animals traversed an alley for 120 trials. The percentage of reinforced trials was either 20, 40, 60, 80, 90, or 100 percent for different groups. The goal box was then placed at the end of one arm of a T-maze and 20 secondary-reinforcement trials were given. Klein found that performance improved as an inverse function of prior reinforcement schedule: Whereas the median number of correct responses for the continuous-reinforcement group was only 11.5 (10.0 was chance performance), the median correct response for the 20-percent group was 18.5.

This finding is somewhat paradoxical in the sense that secondary reinforcing strength diminished with a greater number of reinforcements. However, the continuous-reinforcement group may have experienced frustration during the nonrewarded, secondary reinforcement trials (i.e., the previous goal box became aversive) and therefore switched their responding to the neutral box. In contrast, the partial-reinforcement groups had been reinforced for running in the presence of nonrewarded aftereffects and thus they continued to respond correctly to the former goal box.

Magnitude of Primary Reward

Several studies have indicated that the strength of secondary reward is related to the amount of primary reward (D'Amato, 1955; Butter & Thomas, 1958). For example, D'Amato trained rats to make a correct turn in a T-maze. The reward was one of two goal boxes in which the subjects had been previously fed. The subjects had received 5 food pellets in one goal box and only 1 pellet in the other. D'Amato found that learning was much better when the subjects were allowed to enter the box associated with higher reward.

Delay of Primary Reward

Because the association of the secondary cue with primary reward is central to the development of secondary reinforcement, delays in the presentation of the primary reward should weaken the reinforcing strength of the conditioned cue. Such an effect was shown in two early studies by Bersh (1951) and Jenkins (1950). In both experiments, a buzzer was followed by food after a temporal delay. During a subsequent test, a lever was inserted into the cage and the buzzer was sounded following a response; the number of lever-responses was taken as the index of secondary reinforcement. Bersh found that the optimum delay was between 0 and 2 seconds; i.e., the number of lever-responses declined when the delay had been either 0 or greater than 2 seconds. In the Jenkins study, a larger range of delays, 1, 3, 9, 27,

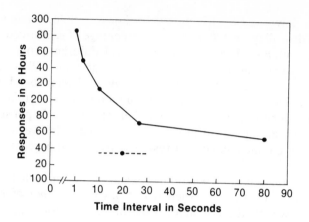

Figure 8–5. Responses on the test as a function of the prior interval between the secondary stimuli and the primary reward.

and 81 seconds, was studied. As shown in Figure 8-5, Jenkins found that the number of responses for the buzzer during a 6-hour test decreased as a function of the original CS-reward interval. Therefore, secondary reinforcement was weakened as the interval between the cue and reward became longer.

Schedule of Secondary Cue Presentation

Finally, a variable influencing the effectiveness of a secondary reinforcer is its schedule of presentation on the test. Generally, responding to gain access to a secondary reinforcer alone is greatly prolonged if the cue is only presented intermittently (Fox & King, 1961; Tombaugh, 1970; Zimmerman, 1957, 1959). For example, Tombaugh (1970) allowed rats to press a lever to obtain sucrose water; 10 trials were given per day for 14 days. The group designated as 100 received reward on every trial; during extinction, this group continued to receive the click of the feeding mechanism (secondary reinforcer), but no primary reward was presented. Group 50-100 received sucrose reward on only 50 percent of the acquisition trials, but the click was sounded on all trials during both acquisition and extinction. Group 50-50 received sucrose reward on 50 percent of the acquisition trials and the click only on those trials. During extinction, this group continued to receive the click on 50 percent of the trials.

Tombaugh used the latency of the lever-press during extinction as a measure of secondary reinforcement. As shown in Figure 8-6, group 50-50 continued to respond faster during extinction than groups 50-100 or 100. This finding illustrated that responding was enhanced by a

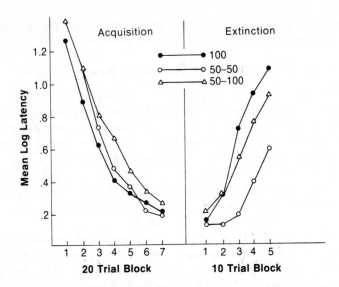

Figure 8–6. Mean latency during acquisition and extinction as a function of the percentage of trials reinforced and the presence of a secondary reinforcement on nonreinforced trials.

secondary reinforcer which was presented intermittently during extinction.

The reason for Tombaugh's finding is not entirely clear; it is likely that the secondary reinforcer strength was greater for group 50-50 because the cue was only associated with reward during acquisition. In contrast, the cue for group 50-100 was followed by both reward and nonreward, and thus was weaker overall.

The finding that intermittent presentation of secondary reinforcers prolongs extinction is of great theoretical importance. As emphasized by Zimmerman (1957), the effects of secondary reinforcement are not necessarily transient. In fact, if techniques such as those just discussed are followed, vigorous behavior may be maintained for an indefinite time by secondary reinforcement alone. Such a possibility is crucial for explaining the maintenance of many behaviors which appear to be virtually autonomous from primary reward.

Theories of Secondary Reinforcement

As discussed in the introduction to this chapter, the nature of secondary reinforcement has been debated for quite a long time, and both classical and instrumental models have been proposed to account for this principle. As with other important principles, there is no clear or simple answer, although considerable progress has been made.

Pavlovian Hypothesis of Secondary Reinforcement

The principal theory, implied in the definition of secondary reinforcement already given, is that the stimulus becomes a reinforcer. In most cases the effects of secondary reinforcers parallel those of primary reward in terms of how behavior is maintained, extinction is prolonged, or new behavior is acquired. Furthermore, it seems that these effects are in direct proportion to the intensity of the primary reinforcer: Secondary reinforcement is stronger when the reward magnitude is greater, when the reward occurs closer in time, or when it follows a greater number of training trials.

In essence, there is an extensive similarity between the conditions which produce a strong secondary reinforcer and those which produce better classical conditioning, i.e., a stronger or more viable CR given to the CS. The secondary reinforcement hypothesis states that a secondary reinforcer is a Pavlovian CS, and, as pointed out by Kimble (1961), the conditions for producing secondary reward are identical to those used in classical conditioning. The fundamental basis for secondary reinforcement, according to this theory, would be the CS-US contiguity. The capacity to modify behavior is, through association, transferred from the US to the CS after repeated pairings.

There is a great deal of evidence to support this theory. One example is the study by Bersh (1951) in which he found that secondary reinforcement strength reached a maximum when the stimulus-food interval was approximately 2 seconds but declined when the interval was 0 or longer than 2 seconds. Here, the function relating secondary reinforcement strength to the CS-US interval resembles functions obtained in classical conditioning. Furthermore, in many of these experiments, the initial training phase in which the secondary reinforcer was established was conducted precisely as in classical conditioning. No instrumental response was specified; rather, the secondary stimulus was simply paired with food without regard to the subjects' behavior. In summary, the view that Pavlovian CS's represent conditioned reinforcers is well supported in the sense that stimuli simply paired with reward are themselves capable of later acting as reinforcers.

Discriminative Stimulus Hypothesis

The discriminative stimulus hypothesis, first formulated by Keller and Schoenfeld (1950), has severely challenged the traditional secondary reinforcement theory. According to Keller and Schoenfeld's position, a stimulus must first be an S_d before it can serve as a secondary reinforcer. Only a discriminative stimulus, one which signals the appropriate time for the response, is capable of becoming a secondary reinforcer. In fact, the secondary reinforcer is the same as the S_d. This hypothesis relies exclusively on the concepts and terms of instrumental conditioning to describe the nature of secondary reinforcers; the conditions used to establish a secondary reinforcer are the methods of instrumental conditioning.

One of the first studies to support this hypothesis was done by Schoenfeld, Antonitis, and Bersh (1950). All rats were allowed to press a lever for food, but the experimental group received a 1-second light only during eating; the control group did not receive the light. Following an extinction period, all subjects were presented with the light following a lever-press. No difference was found in performance between the two groups. Despite the fact that the light had been paired with eating, no secondary-reinforcement effect was shown.

Their explanation (Schoenfeld et al., 1950) for failing to demonstrate secondary reinforcement was to suggest that a stimulus must first be an S_d before being able to function as a secondary reinforcer. This means that the S_d must precede the response of eating, i.e., would signal that eating was appropriate. By virtue of this fact, it would come to elicit the response and thus produce the same effect on behavior as if the response were being reinforced. Without such an ability to elicit conditioned behavior, secondary reinforcing effects are not observed.

Another important study which similarly showed that secondary reinforcers are discriminative cues was done by Dinsmoor (1950), whose rats were trained to press a lever for food in the presence of a light (S_d) and to avoid pressing in its absence. After a total of 200 S_d and S_Δ presentations were given, the subjects were divided into three groups and exposed to extinction conditions. For one group (S_d group) the S_d came on and remained on until the subject made a lever-press. For a second group (secondary reinforcement, or S_r, group), a response produced the S_d, while for a third group (S_Δ group) no stimulus was ever presented.

The mean cumulative responses for the first 50-minute extinction session are shown in Figure 8-7. It is clear that the S_d and the S_r groups did not differ from one another, although both differed from the S control group. Dinsmoor concluded that the discriminative and secondary reinforcing functions of the cue were equal, i.e., secondary reinforcers are discriminative stimuli.

It is difficult to resolve the debate between the secondary-reinforcement and discriminative-stimulus hypotheses because the distinction between an S_d and a Pavlovian CS is not always simple. As discussed in Chapter 3, any stimulus may act as both an S_d and CS,

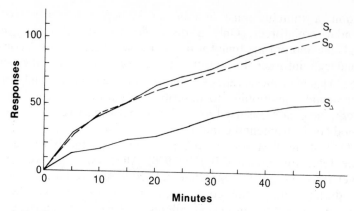

Figure 8–7. Mean cumulative number of responses during the first extinction session for the three groups.

since contiguity with reward also implies the S_d or signaling functions. The fundamental difference with reference to secondary reinforcement, and thus the basic distinction between these hypotheses, is whether the stimulus evokes the response or reinforces it. In both cases, the basic effect on the behavior is the same—maintainance of or prolonging behavior. Experimentally, the difference between these theories is whether the secondary reinforcer must first become an S_d.

There is some evidence to suggest that a stimulus may have secondary reinforcing properties without being an S_d. First, many studies did not require or specify any instrumental response during the training portion of the experiment. Although goal responses, which would be elicited by the stimulus, might facilitate later learning (e.g., lever-pressing) by activating the subject and keeping it in the vicinity of the lever (Wyckoff, Sidowski, & Chambliss, 1958), this has not always been found. Rather, as shown by Crowder, Gill, Hodge, and Nash (1959), a stimulus may become a reinforcer without facilitating behavior through its S_d functions.

Secondly, Ratner (1956) explicitly tested the discriminative-stimulus hypothesis and found that a cue could reinforce learning without being an S_d. Thirsty rats were trained to drink from a dipper when a click sounded. Later a lever was inserted and presses, which produced the click for half the subjects but not the other half, were recorded. Goal approaches to the location of the dipper were also measured. Ratner found that the experimental group that received the click for a lever-press responded much more than the control group. However, there was no difference between the groups in terms of the number of goal approaches. He therefore concluded that a stimulus may be a reinforcer without simultaneously being a S_d.

Thirdly, as pointed out by Hendry (1969, p. 15), the discriminative hypothesis is questioned in a study by Ferster (1953). Using the chaining technique, pigeons pecked a plastic disc during the presence of S_2 in order to produce S_1. During S_1, however, the pigeon was required not to peck. If the subject refrained from pecking for one minute, it was rewarded, otherwise a peck reset the timing mechanism which delayed the presentation of food. Under these conditions, pecking was maintained during S_2 but not S_1. This finding does not support the discriminative stimulus hypothesis, since S_1 was an S_Δ (not an S_d) for pecking, yet it reinforced pecking during the preceding S_2.

Finally, a study by Stein (1958) indicated that a cue may be reinforcing without having discriminative properties. Rats were allowed to press two levers; one produced a 1-second tone but the other did not. The levers were then removed and the tone was paired (400 times) with a small electric current to a specific pleasure center in the brain. (It has been shown repeatedly that such a stimulus, delivered to certain areas of the brain through a thin electrode wire, is positively rewarding.) In the third phase of the study, the levers were reinserted and the number of responses to the tone and no-tone lever was recorded.

Stein (1958) found, as shown in Figure 8-8, that the subjects responded to the lever that produced the tone. Evidently, the tone had

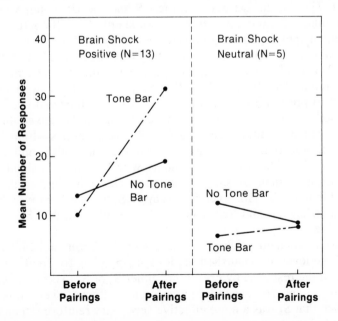

Figure 8–8. Mean number of responses to the two levers (tone and no tone) both before and after the tone-brain reinforcement pairings.

acquired secondary reinforcing properties. Among subjects that got a neutral brain stimulus—neither pleasurable or rewarding nor punishing—no secondary reinforcement effect was observed. This finding challenges the discriminative-stimulus hypothesis because such brain rewards do not involve any form of instrumental response on the part of the subject. The tone could not have been an S_d since no instrumental response was involved, although there was clear evidence for reinforcing effects.

In summary, as Wike (1966, p. 464) has pointed out, discriminative stimuli are able to reinforce behavior, which is not surprising, since such stimuli are followed by reward and thus are also Pavlovian CS's. However, it seems clear that a stimulus may become a secondary reinforcer through association with primary reward without being an S_d. In other words, the fundamental conditions for producing such a reinforcing stimulus are those of classical conditioning.

Information Hypothesis

Wike's conclusion must now be modified because of recent findings. As discussed in Chapter 2, current belief is that a Pavlovian CS acquires excitatory properties not simply through repeated pairings with the US, but rather through its predictive capacity (Rescorla, 1967a). The correlation between the CS and the US is important for determining CS excitatory strength, not simply CS-US contiguity.

When applying these notions to secondary reinforcement, it seems reasonable to conclude that a secondary reinforcer, as a Pavlovian CS, must also be informative. This conclusion was supported by studies by Egger and Miller (1962, 1963). In the first experiment, rats were initially trained to press a lever for food. Next, group A received two stimuli plus food reward, as depicted in Figure 8-9. Group B received the same stimuli, but in addition received free S_1 presentations—followed by neither S_2 nor food—during the intertrial interval. Both S_1 and S_2 were Pavlovian CS's due to their contiguity with food, but S_2 should have been stronger than S_1 since it occurred closer in time to the food reward than did S_1. However, for group A, S_1 was informative whereas S_2 was redundant and, for group B, S_2 was informative while S_1 was unreliable.

After repeated pairings of these stimuli, Egger and Miller reinserted the lever, extinguished the lever-responding for food, and then retrained the response by presenting either S_1 or S_2 to different groups thereby testing for secondary reinforcing effects. There was clear evidence that S_1 was a more effective secondary reinforcer than S_2 for group A, but the reverse was true for group B. The strength of the secondary reinforcer was a function of its informative value: Re-

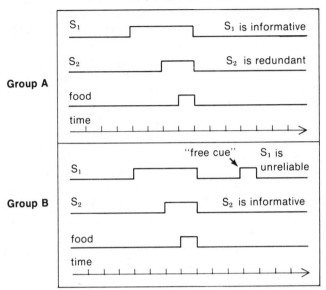

Figure 8–9. Experimental design of the study by Egger and Miller (1962).

dundant or unreliable cues were unable to serve as secondary reinforcers although these results have not been found in all situations (Ayres, 1966).

Egger and Miller's experiments imply that in order for a stimulus to become a secondary reinforcer, it must be informative. Pavlovian CS's, in general, are effective because of their predictive power as discussed in Chapter 2. Therefore, the notion that a secondary reinforcer is a classical CS is supported.

The information value of a stimulus, however, probably doesn't represent the only factor determining the strength of the secondary reinforcer. The information hypothesis, for example, fails to account for experiments such as Jenkins' (1950), in which secondary reinforcement strength varied with the CS-US interval, although the CS (secondary reinforcer) was equally informative for all groups. Secondly, conditioned reinforcement varies with the magnitude of primary reward, even though, from an informative point of view alone, there should be no difference.

Incentive Hypothesis

There is another factor which has been proposed as a general explanation for secondary reward. According to this hypothesis, stimuli which are paired with reward become conditioned motivators, incentives which attract or pull the organism (see Bolles, 1967, and Wike, 1969, for an elaboration of this theory). The stimuli themselves, in the absence of primary reward, elicit fractional goal responses. These responses, in turn, are accompanied by characteristic internal stimuli which enhance response vigor by combining with the other, external discriminative stimuli which elicit the response. The effect is the predicted strengthening of behavior through an increase in the motivation for responding (see Figure 3-6 and the discussion of incentive in Chapter 3).

An example of the motivational effect of an incentive upon performance is found in a study by Marx and Murphy (1961). Rats in the experimental group were first trained to poke their noses into a small hole to obtain food at the sound of a buzzer. A control group received the same training, but the buzzer and food were never paired. In a second phase of the study, all subjects were trained to run down an alley for food. During extinction, the buzzer was presented while the subjects were in the start box on the sixteenth trial and on every fifth trial thereafter.

Results (shown in Figure 8-10) indicated that the experimental subjects ran significantly faster during extinction than the control subjects, while the buzzer had no effect on performance in the control group. Essentially, according to this incentive theory, secondary cues facilitated behavior not by reinforcing the response but by motivating the subject.

While it is reasonable to believe that heightened incentive motivation would facilitate response maintenance and resistance to extinction, it is less clear how incentive motivation would promote the acquisition of new behavior. In this instance, increased motivation should not operate selectively on the to-be-learned response, since the response has not yet been acquired. Rather, it should tend to magnify all performance.

This criticism was noted in a study by Wilkes and Crowder (1960). Hungry rats were first given buzzer-food pairings and were then placed in another cage in which two levers were inserted. A response to one lever produced the 1-second buzzer, while a response to the other did not. Subjects responded to the secondary-reinforcement lever more than to the other lever, indicating that the buzzer had acquired reinforcing properties. The cue could not have facilitated behavior exclusively through incentive motivation. If such had been the case, responding to both levers would have been enhanced.

There is a great deal of work still remaining to determine how secondary reinforcers are established and how they operate to facilitate behavior. It is likely that most cues, in addition to their reinforcing

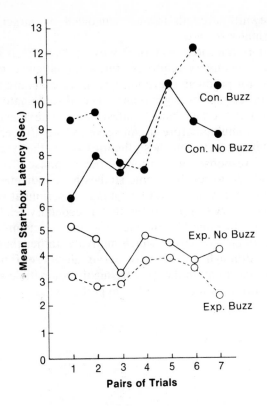

Figure 8–10. Mean start-box latencies during extinction for the experimental and control groups. The "no-buzz" condition represents performance on the trial preceding the "buzz" condition.

properties, are also discriminative stimuli and incentive motivators. At this time, it is not clear how these separate functions interact.

Summary

Secondary reinforcement refers to the acquisition of reinforcing properties by a neutral stimulus due to association with primary reward. The effect may be demonstrated with a chaining technique, in extinction, or in the acquisition of a new response. Secondary reinforcement also has been shown to accrue to stimuli that are associated with shock termination, again using the same three basic techniques.

The strength of a secondary reward is increased by extended training, but not by a higher drive level. A partially rewarded cue is a stronger secondary reward than a continuously rewarded cue, and secondary reward strength is also enhanced by larger and more immediate primary reward.

A secondary reinforcement, or Pavlovian, hypothesis states that a cue becomes a secondary reinforcer through contiguity with reward; the conditions for producing secondary reinforcement are the same as those used in classical conditioning. The discriminative stimulus hypothesis, on the other hand, states that a cue must first be a discriminative stimulus before it can serve as a reinforcer. The basic point of contention between these theories is whether the secondary cue evokes a response or reinforces the response. Although this question is difficult to resolve, some studies have indicated that a cue may be a reinforcer without first becoming a discriminative stimulus.

One other theory has postulated that a secondary reinforcer must be informative, while another, the incentive theory, claims that secondary reinforcers are incentives which motivate, rather than reinforce, the subject. Although a cue may serve all these functions, the Pavlovian and information theories are highly compatible and seem to account for more data than the other hypotheses.

Generalization
Discrimination

Introduction

Stimulus generalization is defined as the process by which organisms perform the CR in response to a novel stimulus—one other than the original CS or S_d—which is similar to the original training stimulus. For example, if the original CS was a tone, the subject would also respond to a novel tone that differed in terms of its pitch or intensity. In fact, virtually any characteristic of a stimulus represents a dimension along which generalization may take place. The response to a generalized stimulus, however, becomes weaker as the stimulus becomes increasingly different from the original training stimulus.

Generalization is an extremely important principle of learning. Without the capacity to generalize, most organisms would be unable to cope with the variety of behavioral demands placed upon them by their complex environment. Organisms would have to relearn appropriate behaviors each time they were confronted with a unique stimulus and, considering the complexity and changing nature of the environment, this would be a formidable task indeed. Generalization, however, provides the basis for responding appropriately to stimuli not previously encountered. If a subject learns to make an adaptive response to one particular signal, the behavior will be elicited by a similar stimulus. The importance of this capacity is especially evident when the stimuli signal potential dangers to the organism.

The perplexing question concerning generalization is: Why does such an effect occur; how do stimuli come to elicit conditioned responses if they were not previously used in the conditioning of those responses? (See reviews by Kalish, 1969; Mednick & Freedman, 1960; Prokasy & Hall, 1963; Terrace, 1966b; and Mostofsky, 1965.)

Methods

There are two basic procedures for demonstrating generalization. The first, the repeated-stimulus technique, involves conditioning a response to a specific stimulus and then, either during continued acquisition or during extinction, presenting a series of varied, but similar, stimuli in a random order to each subject. This technique is often used, even though it has the disadvantage that the response to one generalized stimulus may be severely influenced by responding to other generalized stimuli.

A method which avoids this problem is the single-stimulus technique. During extinction, different groups of subjects receive only one generalized stimulus; the ability of each generalized stimulus to evoke the response is assessed on the basis of group comparisons. The first such extinction response is the most useful in this regard, since continued extinction modifies the response strength and thus the index of generalization.

Examples of Generalization

Primary Stimulus Generalization

When investigating generalization in the laboratory, lights, tones, and other external stimuli are normally used as stimuli. Variations in these stimuli are typically in terms of intensity, frequency, and hue. These easily specified dimensions allow for accurate correlations between strength of response to a generalized stimulus and the amount of change in that stimulus. Generalization based upon variations in such physical dimensions is called primary stimulus generalization; generalization based upon a stimulus dimension acquired through learning is termed secondary generalization.

In one of the most elegant experiments on primary generalization (Guttman & Kalish, 1956), pigeons pecked a plastic disc to obtain food. The disc could be illuminated by a system of lights, and, with the additional use of chromatic filters, the color, or wavelength, could be exactly controlled. For different groups of subjects, the wavelength was either 530, 550, 580, or 600 m$_\mu$. Food could be obtained (on a VI-1 minute schedule) while the disc was illuminated, but not when it was dark. The 60-second S_d periods were alternated with 10-second S_Δ periods.

The generalization test was carried out during extinction. Each group was given eleven different, randomly presented stimuli during extinction, only one of which was the original S_d; the other ten were generalized stimuli which differed from the original S_d by as much as 60 m_μ. The measure of generalization was the number of responses given to each stimulus during extinction. The total number of stimulus presentations was 132, equaling twelve presentations of each stimulus.

The results are illustrated in Figure 9-1. Each group responded most frequently in the presence of their original S_d. Responding to the generalized stimuli decreased as a function of difference between the original S_d and the generalized stimulus. This orderly function is called the generalization gradient, for it shows a graded response pattern. While similar stimuli elicit many responses, responding is virtually absent when the generalized stimulus differs appreciably from the training stimulus.

The basic gradient effect is more clearly shown in Figure 9-2, where responding is plotted according to the degree of difference from the original S_d, measured in m_μ. The mean gradients in Figure 9-2 represent all groups combined; this figure also shows the effect of extinction on generalization, as a gradual reduction in response

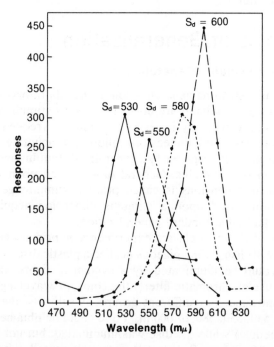

Figure 9–1. Mean total responses given for each generalized stimulus. Curves illustrate mean generalization gradients for the different groups.

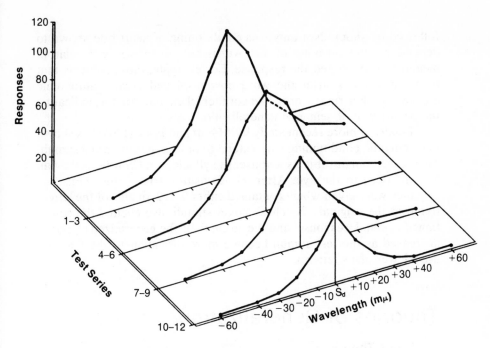

Figure 9–2. Mean generalization gradients as a function of blocks of extinction trials and difference between the generalized and training stimuli.

strength during extinction, which is manifest in the overall height of the gradient.

Semantic Generalization

The foremost example of secondary-stimulus generalization is semantic generalization, in which the meaning of a stimulus is the dimension on which the generalization is based. First a response is conditioned to a word and then a generalized response is obtained when other words—synonyms or homonyms—or the object to which the word refers, are given as stimuli. In one early study Razran (1939) conditioned a salivary response in humans by presenting such words as urn (CS) followed by candies (US). During a generalization test, Razran presented vase (a synonym) or earn (a homonym) and discovered that the subjects salivated to those words as well; thus, a generalized response was made based upon the meaning or sound of the word.

Another example of this phenomenon was provided in the important study by Lacey, Smith, and Green (1955) mentioned in Chapter 2. Subjects gave associations to a list of words, some of which were followed by shock. Not only was conditioning of heart rate shown to depend on the meaning of the word, but other words, similar in meaning, also elicited the response. For example, those subjects that received a shock after the word cow displayed greater autonomic arousal to other rural words like corn, chicken, and tractor, indicating the occurrence of semantic generalization.

Finally, a more recent study by Abbott and Price (1964) has shown that semantic conditioning may extend to other types of verbal stimuli. They gave human subjects a nonsense syllable as a CS followed by an air puff (US) to the eye. After 90 acquisition trials, other nonsense syllables were given which contained either 0, 1, 2, or all 3 of the letters found in the original CS. These novel stimuli also elicited the conditioned eyeblink response, and the strength of the generalized response decreased as the test stimuli became more dissimilar to the CS (i.e., used fewer of its letters).

Theories of Generalization

Hull's Theory

Hull (1943) suggested that generalization was a primary process, representing an inherent, fundamental process in learning, and not a secondary by-product of some other process. According to Hull, during instrumental conditioning, the response actually became conditioned to a "zone" of stimuli rather than to one simple stimulus. The zone was structured in terms of the sensory similarity of the stimuli. During a generalization test, the response could be elicited by all the stimuli contained in the zone, although more dissimilar stimuli were weaker in their ability to evoke the CR. Generalization had occurred because varied, but similar, stimuli were actually conditioned to the response during acquisition, and the generalization test merely provided an opportunity for these stimuli to evoke the conditioned response.

Hull's theory had some trouble in accounting for the fact that generalization gradients are rarely symmetrical; that is, more intense generalized stimuli usually evoke larger responses than less intense generalized stimuli, even though they differ from the original CS by the same amount. In a later paper, Hull (1949) employed an additional concept which strengthened his thesis and accounted for the fact that response strength tends to vary according to CS intensity.

Hull's basic notion was supported in a series of studies by Hovland (e.g., Hovland, 1937a,b). In the first experiment, four tones (153, 468, 1000, or 1967 Hz) were equated for loudness. One was then used as the CS in the conditioning of a GSR response. During extinction, all of the tones were presented and response strength was measured in terms of GSR amplitude. Hovland found that generalized responses were given, and that their magnitude decreased as a function of the dissimilarity between the generalized tone and the original CS.

The principal reason that Hovland's (1937a) experiment supports Hull's position is that the four tones were widely separated (25 j.n.d.'s apart) and thus could not have easily been confused. In fact, the frequencies of tones were initially chosen because judges rated them as highly discriminable. The fact that generalized responses were given, therefore, could not be attributed to the possibility that the subjects were unable to distinguish between the generalized tones and the CS on the generalization test.

Lashley-Wade Hypothesis

The main alternative to Hull's theory, which Lashley and Wade (1946) formulated, was that generalization occurs simply because subjects fail to discriminate between stimuli. According to this position, subjects are confused during the generalization test, and therefore generalization results from imperfect discrimination. Furthermore, the stimulus dimension, along which generalization takes place, is not known to the subject until the generalization test. Therefore, it was incorrect to postulate that a zone of stimuli had been conditioned during acquisition as previously claimed by Hull.

A study which supports this hypothesis was performed by Kalish (1958). Humans viewed a patch of colored light for one minute, after which they were briefly presented with different hues and told to release a telegraph key if they thought that the hue matched the first one. The initial color was either 500, 530, 560, or 580 m_μ, and the four generalized test stimuli varied above and below those values in 10 m_μ intervals.

Kalish (1958) found that the generalization function matched the discrimination function; he could predict the degree of generalization on the basis of previous information showing how easily humans can discriminate colors. When the colors were quite dissimilar and could be easily discriminated, a steep generalization gradient was obtained; but when discrimination was poor, a great deal of generalization (flat gradient) was found. More recently, Thomas and Mitchell (1962) have confirmed this finding, although other studies have failed to show such a close, inverse correspondence between discrimination and generalization (e.g., Guttman & Kalish, 1956).

Further support for the Lashley-Wade hypothesis is found in the fact that when generalization is tested by presenting several different stimuli during extinction, the gradient becomes progressively steeper with continued extinction trials (e.g., Kalish & Haber, 1963; Friedman & Guttman, 1965). The plausible explanation for this finding is that subjects develop a basis for discriminating between the stimuli during extinction. They become less confused after some experience with all the stimuli and therefore respond progressively less to stimuli that are dissimilar from the original CS.

This effect was clearly shown in Friedman and Guttman's (1965) experiment. Pigeons, trained to peck a colored disc for food, were given an extinction test and their responding to a number of generalized stimuli was recorded. As shown in Figure 9-3, the generalization gradient was relatively flat during the first quarter of the extinction training but got progressively steeper during the later quarters.

Mediated Generalization

A third hypothesis, presented by Osgood (1953) to explain generalization, accounts particularly well for semantic generalization, although it is not limited to that example. According to Osgood, different stimuli may become more or less equivalent (i.e., evoke the same CR as in generalization) if they elicit the same mediating response. Such a response is an internal, intermediary step in the overall chain of events starting with the stimulus and ending with the CR. Early perceptual learning produces these implicit, mediating responses. Clearly, physically similar stimuli will elicit the same mediating response more so than dissimilar stimuli.

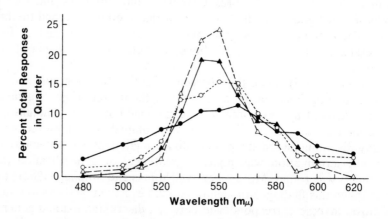

Figure 9-3. Mean generalization gradients during the first (filled circles), second (open circles), third (filled triangles), and fourth (open triangles) quarters of responding.

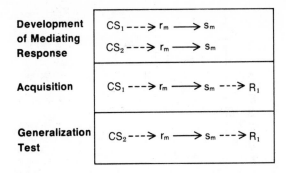

Figure 9-4. Schematic matrix illustrating the existence of the same mediating response (r_m) to different stimuli (CS_1 and CS_2), acquisition of a response (R) to one stimulus, and a generalized response by the second stimulus via elicitation of the mediating response.

The mediation hypothesis is illustrated in Figure 9-4. Both stimuli evoke the same implicit response (r_m), which, in turn, is accompanied by an internal, mediating stimulus (s_m). During conditioning, the implicit stimulus is the final link in the chain prior to the instrumental response (R). During generalization, the response is elicited again by the mediating stimulus because the generalized stimulus had also elicited the implicit, mediating response.

A number of studies supported this notion (e.g., Grice & Davis, 1958, 1960; Grice, 1965). For example, Grice and Davis (1960) used three tones as stimuli (240, 850, and 1900 Hz). Subjects were instructed to respond to the middle tone plus one of the others by pushing a small lever with the right hand. For the third tone, they were told to push a different lever with the left hand. An air puff to the eye followed the middle tone only, making that tone the CS for the conditioned eyelid response. Eyeblink responses were recorded during the last 60 trials in response to all three tones.

The results indicated that eyeblink responses were elicited by both of the generalized stimuli; thus, generalization did occur. However, as shown in Figure 9-5, the gradient was not symmetrical. The tone, for which the required lever-response was the same as the middle tone (both right-hand responses), elicited more generalized eyeblink responses than the other tone for which a left-hand lever-response was required. Grice and Davis' (1960) study confirms the mediation theory by showing that when stimuli elicit the same mediating response (developed in this study by having the subjects push a lever) the stimuli are more likely to elicit the same CR than when the mediating responses differ. Therefore, the important fact which determines the

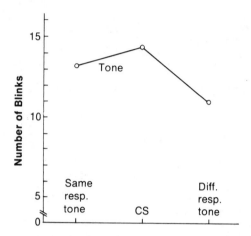

Figure 9–5. Mean number of eye-blink responses to the CS and two generalized tones. One generalized tone required the same lever response as the CS while the other employed a different response.

extent of generalization, according to this theory, is the similarity of the mediating response rather than simply the physical similarity between the CS and the generalized stimulus.

Factors Influencing Generalization

Mediating responses undoubtedly exist and are important in determining generalization. However, because the nature of these responses is postulated and is not objectively specified, there is no means of measuring them. In contrast, the issues surrounding the Hull and Lashley-Wade controversy have been investigated more successfully. Most of the results in the following discussion support the Lashley-Wade position, although several generally favor Hull's formulation.

Number of Training Trials

Margolius (1955) trained rats to obtain food by opening a door that was located in the center of a 79 cm² white circle. Different groups received either 4, 16, 64, or 104 training trials. Each group was then subdivided and given 30 extinction trials in which the circle was either 79, 63, 50, or 20 cm². Generalization was assessed by measuring the total number of responses, as well as their latency, made to each stimulus.

Margolius found that the speed and frequency of response generally increased with greater training. Moreover, the slope of the generalization gradient became steeper with more training. This latter finding supported the Lashley-Wade hypothesis, since it indicated that less generalization occurred with increased training, which provided more opportunity for discrimination. This finding has been obtained by most investigators (e.g., Hearst & Koresko, 1968) although not by all (e.g., Jensen & Cotton, 1961). A larger number of studies have indicated that the generalization gradient does become steeper with continued extinction as well (e.g., Wickens, Schroder, & Snide, 1954; Kalish & Haber, 1963; Friedman & Guttman, 1965), as illustrated in Figure 9-3.

Training-Test Interval

The interval between the training and generalization phases influences the degree of generalization; generally the gradient becomes flatter with time, indicating greater generalization. For example, Perkins and Weyant (1958) trained rats to traverse a straight alley to obtain food on an intermittent schedule. Following training, subjects were given 18 nonreinforced test trials. Two groups received these test trials in the same training alley, while two other groups were tested in a novel alley which differed in color. In addition, one group in each of these two conditions received test trials immediately after their last training trial; the other groups received their trials after a 7-day interval. Thus, the four groups were identified as: immediate test–same alley; immediate test–novel alley; delayed test–same alley; and delayed test–novel alley.

The results are shown in Figure 9-6. The two immediate test groups show the predicted generalization effect: Responding was very fast when the immediate test was given using the same alley, but very slow when using the novel alley. The generalization gradient was quite steep, reflecting the relative inability of the novel stimulus to elicit the response. In contrast, the delayed groups showed a different pattern. Most striking was the fact that speed was greater in the novel alley than it had been for the group tested immediately. Thus, the generalization gradient had flattened over time.

Perkins and Weyant (1958) argued that the subjects had forgotten the apparatus color during the 7-day delay period but not the general response of running in the alley. Therefore, responding for the delayed-novel group was not disrupted by the introduction of a novel cue as it had been for the immediate-novel group. A flattening of the gradient of generalization to fear stimuli has also been obtained after delays of only 24 hours or less (e.g., McAllister & McAllister, 1963;

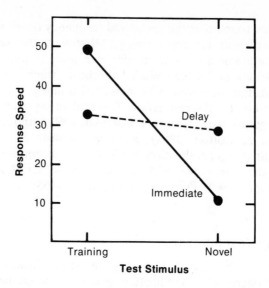

Figure 9–6. Median running speed as a function of the training or novel alley for the immediate and delayed groups.

Desiderato, Butler, & Meyer, 1966; Desiderato & Wasserman, 1967), suggesting that fear of such generalized stimuli may actually increase over time. However, Thomas and Lopez (1962) also found the effect, within 24 hours, for an appetitive task, thus supporting Perkins and Weyant's notion of differential forgetting. In general, these observations tend to favor the Lashley-Wade hypothesis, since forgetting the apparatus stimuli but not the basic response implies a reduction in discrimination and an increase in generalization.

Early Experience

An important finding, which also supports the Lashley-Wade theory, was shown by Peterson (1962). He raised ducks, from birth, in a monochromatic light—an environment in which all available light was of a single wavelength (589 m$_\mu$). This procedure completely denied the experimental group any experience with different colors and thus any basis for future color discrimination. Control animals were raised under normal lighting conditions. Second, all subjects were taught to peck a disc which was illuminated with the same monochromatic light (589 m$_\mu$). In the third phase of the study, a generalization test was administered in which the color of the disc varied from 490 to 650 m$_\mu$.

Peterson (1962) found that the typical generalization gradient was displayed by the control group. Rate of responding to the disc decreased as a function of the dissimilarity between the generalized stimuli and the original training stimulus. In contrast, the experimental group displayed complete generalization, i.e., responded equally to all the test stimuli. Peterson suggested the cause was their inability to discriminate colors. Similar results for dark-reared monkeys were found by Ganz and Riesen (1962).

In a second experiment, Peterson (1962) showed that the experimental ducks could learn to respond differentially to different colors if they were appropriately reinforced. This fact argues against the criticism that his earlier findings had been due to a lack of visual acuity on the part of the experimental subjects. However, the issue is still not totally resolved, since there may be some innate color preferences in birds which could have biased Peterson's results (Tracy, 1970).

Prior Discrimination

A result which supports the Lashley-Wade position regards the effect of prior discrimination training on subsequent generalization. In general, discrimination training—reinforcing the subject for responding to the S_d, and withholding reward for a response to the $S_\triangle$—results in a steeper generalization gradient later when both the S_d and $S_\triangle$, as well as other generalized stimuli, are presented to the subject.

These results have been found by numerous investigators (e.g., Hanson, 1959, 1961; Jenkins & Harrison, 1960; and Thomas, 1962). For example, in the Hanson (1961) experiment, pigeons pecked a disc, illuminated with monochromatic light of 550 m_μ, to obtain food. The experimental group was rewarded for a response during the 550-m stimulus but not during a 540- or 560-m_μ stimulus. The control group never experienced the two $S_\triangle$'s and thus did not receive discrimination training. In the final phase, both groups were given a generalization test using stimuli ranging from 490 to 610 m_μ.

As seen in Figure 9-7, orderly gradients were obtained for both groups; however, the gradient for the experimental group was much steeper than that for the control group. This finding indicates that when subjects are explicitly trained to differentiate between stimuli, there is little generalization. Actually, prior discrimination along any dimension sharpens subsequent generalization. The assumption is that a similar discrimination process or lack thereof accounts for the generalization gradient in subjects who were not explicitly trained to discriminate.

An interesting finding, related to discrimination training and subsequent generalization, is termed the peak shift (see Bloomfield, 1969, for a discussion of this phenomenon). As first shown by Hanson (1959), the peak shift occurs on a generalization test following discrimination training when there is a shift in the distribution of

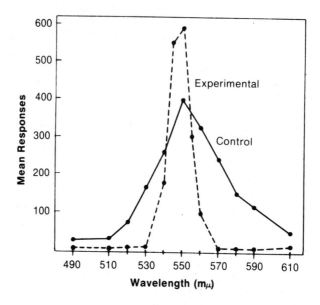

Figure 9–7. Mean responses for the discriminative (experimental) and nondiscriminative (control) groups as a function of stimulus wavelength during the generalization test.

responses along the generalization gradient so that maximum responding occurs not to the S_d, but to a stimulus that lies on the side opposite the $S_\triangle$. For example, in Hanson's study, pigeons pecked a disc that was illuminated with monochromatic light (550 m$_\mu$) to obtain food. Different groups received $S_\triangle$'s of 555, 560, 570, or 590 m$_\mu$; the control group didn't receive any $S_\triangle$. The generalization test for all subjects consisted of thirteen different stimuli ranging from 480 to 620 m$_\mu$.

Hanson found that the point of maximum responding had shifted for each experimental group away from the S_d (550 m$_\mu$). As shown in Figure 9-8, the shift along the generalization dimension was to the side opposite the $S_\triangle$; the peak of responding had shifted. Furthermore, the magnitude of the shift was an inverse function of the S_d-$S_\triangle$ difference.

The exact cause for this phenomenon is not entirely clear, although the peak shift does depend upon prior discrimination training. When special training procedures are used which result in errorless discrimination training, the peak shift does not occur (Terrace, 1963). This fact led Terrace to hypothesize that because of nonreinforcement the $S_\triangle$ becomes an aversive stimulus during discrimination training and that the peak shift represents an emotional reaction, an overshooting on the part of the subjects to avoid the aversive $S_\triangle$. However, when no errors are made, the $S_\triangle$ does not become aversive and no peak shift is

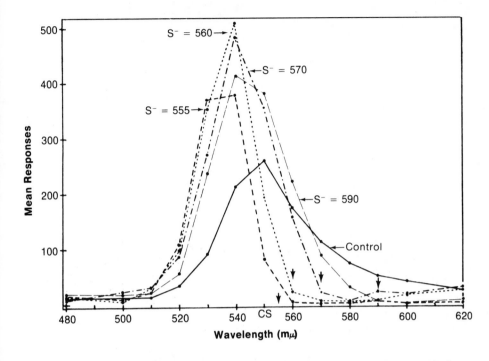

Figure 9–8. Mean responses for the control and experimental (discrimination) groups as a function of stimulus wavelengths during the generalization test. Arrows indicate the positions of the negative stimuli for the experimental groups.

found. In further support of this point, Terrace (1966a) found that the peak shift disappeared after extended discrimination training during which the emotional reaction declined.

Motivation

One final factor in generalization, which supports Hull's theory, regards the effect of motivation. In general, increased hunger during the generalization test produces a steeper gradient (e.g., Coate, 1964; Rosenbaum, 1953; and Thomas & King, 1959). This effect cannot easily be explained on the basis of a decrease in discrimination, since the discriminability of the stimuli should be the same for both high- and low-drive groups.

Kalish and Haber (1965), for example, trained three groups of pigeons to peck a colored disc (550 mμ) for food. The groups differed in terms of their food-deprivation level—their weight was restricted to 90, 80, or 70 percent of their normal weight. Because the generalization-test stimuli varied from 490 to 550 mμ only, the authors investigated one

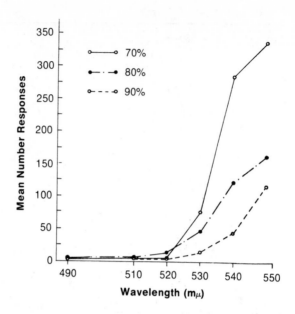

Figure 9–9. Mean number of responses to the generalized stimuli as a function of deprivation level (percentage of normal weight).

side of the generalization gradient. As shown in Figure 9-9, the slope of the gradient varied as a function of deprivation level. Whereas the low-deprivation (90 percent) group displayed a relatively flat gradient, the high-deprivation (70 percent) group's gradient was quite steep.

In summary, the evidence seems to favor the Lashley-Wade position, that generalization is related to the subject's relative failure to discriminate. However, in manipulating many of the variables mentioned above, changes in both the slope as well as the overall height of the gradients have been observed, suggesting that the solution to these problems is very complex. Furthermore, the functional stimulus dimensions along which subjects generalize may be somewhat different or more inclusive than suspected by the experimenter, and thus may be obscured in experiments in which colors or tones are used exclusively. Certainly, the work on semantic generalization shows the complexity and variety of generalization dimensions.

Discrimination

Discrimination is closely related to generalization, but is an area of study in its own right. Operationally, discrimination training consists of reinforcing a response to the S_d but not to the S_Δ. Actually, in many cases, two S_d's are employed but the reinforcement schedules associated with each stimulus differ. The general result is differential responding to these stimuli (see Gilbert & Sutherland, 1969, for a more complete discussion of discrimination).

Basic Methods

There are several different methods for conducting discrimination training. The varied results, based upon which of these methods is used, in turn, have a bearing on the kind of process which is inferred to take place during discrimination. The first method is simultaneous discrimination, in which both the S_d and $S_\triangle$ are presented to the subject at the same time, and the subject must decide which stimulus is the S_d. For example, two differently colored discs may be located in a pigeon's chamber. A response (peck) to the S_d would produce food, while a response to the $S_\triangle$ would not. Another example would be a T-maze in which the animal must choose the correct turn.

A second technique is called successive discrimination. In this case, one of the stimuli is presented first and the subject responds or not depending on whether it is the S_d or $S_\triangle$. Next, the second stimulus is presented. No choice is made between the stimuli; rather, the subjects gradually learn to respond in the presence of the S_d but to withhold responding in the presence of the $S_\triangle$.

Theories and Issues in Discrimination

The traditional continuity theory, stemming from the work of Hull (1943) and Spence (1936), has formed the background for most of the research on discrimination. This position states that discrimination— differential responding—occurs because the correct response is reinforced, while the incorrect response is not reinforced. More specifically, a reward leads to an increment in excitatory strength for the S_d, while nonreward increases the inhibitory strength of the $S_\triangle$. Both of these tendencies generalize to other similar stimuli and the gradients summate to produce the differential responding.

To clarify by example, the excitatory S_d might be one color (e.g., 550 m_μ) while the inhibitory $S_\triangle$ might be another (560 m_μ). However, the excitatory and inhibitory tendencies generalize, so that the 550 m_μ stimulus would also be somewhat inhibitory and the 560 m_μ stimulus somewhat excitatory. According to the theory then, the net effect for any two stimuli, i.e., the extent of differential responding, would simply be the summation of the positive and negative tendencies.

Transposition

A number of findings have challenged the Hull-Spence position. One of the most notable is transposition, i.e., the tendency for subjects to respond to the relationship between stimuli rather than their absolute characteristics. This effect was first found by Kinnaman (1902) and has been observed many times since (see Hebert & Krantz, 1965, for a review).

As an example of transposition, a subject is trained to respond to a 200 cm² stimulus but not to a 150 cm². On a subsequent test, the 200 cm² S_d is presented along with a novel stimulus of 250 cm². The subject then begins to respond to this new stimulus, not the original S_d. The new stimulus is the larger of the two, just as the original S_d had been the larger during the initial discrimination. This effect is paradoxical since the subject was never reinforced for responding to this new stimulus. Thus, the shift to the novel stimulus is made on the basis of the relationship between the original stimuli.

Spence (1937) attempted to incorporate this phenomenon into his original theoretical framework first by assuming that the excitatory and inhibitory gradients summated, and then by assigning values to each gradient so that he could predict the transposition outcome. According to this formulation, transposition occurs when the net effect is greater for the novel stimulus than for the original S_d. In other words, the inhibitory tendency for the novel stimulus is a great deal less because of its distance from the S_Δ. Its excitatory strength, however, is only slightly reduced since it is relatively close to the S_d on the generalization dimension. In contrast, the S_d itself, although having the largest excitatory value, is close to the S_Δ and therefore also has a large inhibitory component. These outcomes depend entirely upon the shape of the gradients. Some support has been shown for Spence's formulation (e.g., Ehrenfreund, 1952), although other investigators have failed to confirm it (Lawrence & DeRivera, 1954). In conclusion, it appears that subjects are able to respond both in terms of relational as well as absolute stimulus characteristics. More work is needed, however, to establish the exact conditions which determine the response.

Learning Sets

One of the most important phenomena in discrimination learning is the learning set (see Medin, 1972; Miles, 1965; and Reese, 1964, for reviews). The basic technique involves the successive solution of problems by making a choice between an S_d and S_Δ. With enough training, subjects not only learn to solve each problem, but they get progressively better at doing so. At first it takes quite a few trials before the subject correctly responds to the S_d; however, after extended practice with a variety of stimuli, solutions are acquired by the subjects in only a few trials. Thus, subjects appear to learn how to learn.

The classic study on learning sets was done by Harlow (1949). He used the Wisconsin General Test Apparatus, consisting of a tray on which objects (e.g., blocks and other geometric shapes of various colors) were placed. Two objects were used at a time and food was located under one (the S_d). The tray was positioned in front of a rhesus monkey whose task was to obtain food by choosing the correct object. A number of trials were given while varying the position of the stimulus objects on the tray; the measure of mastery was the percentage of correct choices. This procedure was then repeated with two different objects for a total of 344 different problems.

As the monkeys were given more and more problems to solve, their ability to master a given problem improved, as shown in Figure 9-10. For the first eight problems, learning was relatively slow and after six trials mean percent correct was only about 75 percent. However,

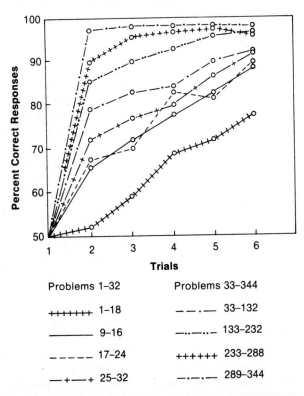

Problems 1–32		Problems 33–344	
++++++	1–18	— — . —	33–132
————	9–16	—..—..—	133–232
– – – –	17–24	++++++	233–288
—+——+	25–32	—.—.—	289–344

Figure 9–10. Mean percent correct responses on the first six trials of a problem as a function of groups of problems as indicated in the key.

with each successive group of problems, acquisition rate improved. On the last block of problems, the monkeys required only one trial before they solved the problem, and chose the correct object on approximately 97 percent of the trials.

Harlow's work is extremely important for a number of reasons. First, studies of learning sets illustrate what, in fact, occurs in an organism's normal life where the solutions to each problem are influenced by previous solutions or partial solutions. This is particularly true, of course, for humans, who can profit more than many other species from previous learning experiences. Therefore, a list of basic factors influencing learning must include more than the immediate variables manipulated over a relatively short time for an experimentally naive subject. The list must certainly include the longer-term effects of solving many problems if learning theory is to account accurately for normal, complex behavior. Learning-set techniques provide a useful means for studying these longer-term effects and of integrating them into the larger learning-theory framework.

Second, learning-set techniques have provided a valuable means of comparing learning abilities across many species of animals. It is difficult to establish whether, say, a dog is smarter than a rat for the simple reason that they are so physically different. Each species is limited with respect to the kinds of responses it is physically capable of performing and the kinds of sensory stimuli with which it is capable of dealing. In general, the speed at which a learning set develops is positively related to the phylogenetic level of a given species. Learning sets have been investigated in a wide range of animals including fish, birds, rats, cats, raccoons, horses, a variety of nonhuman primates, and humans. For example, Warren (1965) summarized some previous work by other investigators which illustrated this principle. As shown in Figure 9-11, all of the species tested improved in their ability to learn problems, as measured by their performance on the second trial of a given problem. The rate at which this took place, and the extent of improvement, varied greatly as a function of the species. In fact, a substantial difference was found even between squirrel monkeys and rhesus monkeys. Learning-set performance is certainly not an immutable index of intelligence, but it does help to establish relative learning abilities. Furthermore, it can be used successively to investigate differences within a species as a function of such conditions as brain damage, retardation, and other dysfunctions.

A great deal of research has been done on learning sets in terms of identifying the effect of certain variables on speed of acquisition. For example, the magnitude of reward is positively related to speed of learning-set acquisition (Schrier, 1958), although the degree of food deprivation is not (Miles, 1959). Second, Levine, Levinson, and

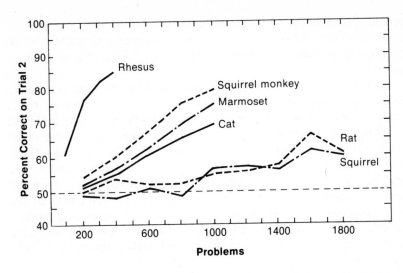

Figure 9–11.   Mean percent correct response on trial two as a function of blocks of problems for various species.

Harlow (1959) have shown that asymptotic performance is not a function of the number of problems encountered but rather the total number of trials. After 64 days of training they compared the performance of a group of monkeys that received daily training of 12 trials per problem (3 problems) with subjects that received 3 trials per problem (12 problems) and found no difference. Third, the rate at which learning sets are formed depends on the type of stimulus objects used in training. Three-dimensional stimulus patterns are superior in this regard to two-dimensional patterns (Harlow & Warren, 1952). Fourth, it has been shown that animals can retain their ability to make discriminations over long intervals (e.g., Mason, Blazek, & Harlow, 1956; Riopelle & Moon, 1968; Strong, 1959). Finally, several investigators have shown that acquisition of one type of learning set (e.g., using patterns) improves subsequent acquisition of another type (e.g., using objects; Harlow & Warren, 1953). However, other studies have shown that a learning set based upon position discrimination can interfere with subsequent acquisition of a learning set based on object discrimination (e.g., Warren, 1959).

The traditional theory of learning would account for learning sets by claiming that they occur through a combination of reinforcement for the correct response and inhibition of the incorrect response. However, Harlow (1950, 1959) challenged that viewpoint. His error-factor theory is a uniprocess one in the sense that it accounts for learning-set formation with a single principle. Namely, learning sets are formed because the subject learns what not to do: The correct response is present from the start but it competes with other inappropriate responses. The subject simply learns to avoid making those incorrect responses.

Harlow proposed this hypothesis because he observed systematic error patterns during the formation of the learning set. For example, subjects perseverated (or repetitively chose an incorrect object) based on either innate or learned object preferences; some subjects displayed a strong position preference without regard to the object itself. These error tendencies decreased as the learning set became stronger. The error patterns suggested to Harlow that the traditional view, based on simple strengthening of the correct response and inhibition of the wrong response, was incorrect. If this traditional theory was correct, the error pattern should have been essentially random. However, the errors were not committed randomly but rather according to systematic patterns.

A third account of learning-set formation was given by Levine (1959, 1965). He proposed that the subject develops hypotheses, or strategies which get reinforced, and, in turn, account for both the patterns of errors as well as the ultimate improvement in learning ability. For example, one strategy, or hypothesis, is position preference; another is win-stay with object, lose-shift to the other object. The subject responds long enough with a given strategy to learn if it works. If not, the subject selects another until he chooses one that does work. Levine's theory accounts for much of the data although there are problems still to be resolved (Medin, 1972).

Summary

Stimulus generalization is observed when a behavior is performed in response to a novel stimulus that is similar to the original training stimulus. Generalization may be assessed by presenting to the subjects several generalized stimuli (repeated-stimulus technique) or only one generalized stimulus (single-stimulus technique). While most experiments on generalization involve a physical dimension such as intensity or wavelength, a generalized response may also be made to a novel word or picture based upon the meaning of that word or picture. This type of responding is an example of semantic generalization.

Hull theorized that gene1tlized stimuli were actually conditioned, while the Lashley-Wade position claimed that generalization occurred because of confusion between the generalized and original stimuli. A third hypothesis stated that generalized responses were given only when the stimuli elicited the same mediating response.

With regard to the variables affecting generalization, it is clear that the generalization gradient becomes steeper with extended training but generally flatter with time. Lack of early experience with a physical dimension, such as color, produces complete generalization. Prior discrimination and increased drive steepen the generalization gradient. In addition, discrimination training prior to the generalization test may cause a peak shift on a subsequent generalization test.

The traditional theory of discrimination states that the S_d becomes excitatory while the S_Δ becomes inhibitory. The phenomenon of transposition, however, shows that the responding to novel stimuli may reflect the relationship between the original S_d and S_Δ rather than the absolute characteristics of the S_d.

Another important phenomenon, learning sets, refers to the progressive improvement in problem solving, i.e., learning to learn. This phenomenon, too, challenges the traditional theory of discrimination since systematic error patterns, rather than random errors, are found during the formation of learning sets.

Reward:
Theory & Application

Introduction

In Chapter 2, contiguity, or CS-US correlation, was cited as the fundamental law governing classical conditioning. For instrumental conditioning (Chapter 3), reinforcement contingency was offered as the basic principle. However, the way in which reward operates—its mechanism of action—was not specified.

Various attempts have been made to understand how or why reward works to strengthen learned behavior. Some of these use abstract or theoretical concepts, others descriptive terminology, while a few have discussed reward in mathematical or biological terms. No simple resolution exists for this exceedingly complex issue, and it is probable that none will be offered in the near future because current theories of learning tend to cover only a limited range of phenomena.

A theory in general is a system for relating facts to one another. Theory doesn't really explain the nature of reward in the sense of presenting a physical reality, although certainly physical facts are involved in establishing theoretical principles. Rather, theory accounts for observed relationships within a particular linguistic context (e.g., mathematical, biological, conceptual). Even more important, theory represents a strategy for research, a direction for further study based upon deductions from the relationships which it establishes.

It is not surprising, then, that no simple resolution has been achieved regarding how or why reward works. What has been accomplished represents a variety of viewpoints, different modes of analysis, and different linguistic treatments of behavioral phenomena. No comprehensive coverage of reward theories is provided here; instead, the chapter focuses on some major viewpoints and the utility and future development of learning theory (see Wilcoxon, 1969, for a review of the history of reinforcement theory).

Drive Reduction Theory

Drive reduction was suggested by Hull (1943) as the primary reward mechanism (see Bolles, 1967). Drive is defined as a state of the organism which arises from a biological need like hunger. In itself, drive is unobservable; it is defined in terms of the operations used to produce the need, such as hours of food deprivation (see Weinstock, 1972). In essence, the theory states that stimuli that reduce drive are reinforcing, and thus they produce an increment in response strength.

Hull actually made very little distinction between need and drive. Later formulations, however, pointed out that they were independent concepts, in part because drives could be learned. The clearest example, of course, is acquired fear in which case the animal is motivated or has drive. Yet fear is not dependent upon a physical imbalance except during conditioning where pain may cause tissue damage; it is a secondary, or learned, source of motivation. This implies that hunger, a primary drive, could be conditioned as well. Conditioned hunger would then represent a source of drive in the absence of a biological imbalance (see Cravens & Renner, 1970, for a review).

A second argument for need-drive independence is the fact that need is usually alleviated long after the subject has finished responding. This suggests that the organism's reward would be delayed a considerable length of time if need reduction was the basic mechanism of reward. Because learning is achieved rapidly, it is unlikely that need reduction alone is responsible.

However, it has been shown that a reduction in need will produce learning. For example, Coppock and Chambers (1954) used hungry rats that were restrained in a small cage such that they could only move their heads to either side to interrupt a photoelectric beam. If they oriented their heads to one side, but not the other, they were rewarded with a glucose solution which was injected directly into their veins. A control group was given normal saline in the same manner. As shown in Figure 10-1, the glucose group learned to maintain the correct position, but the saline control group did not. Coppock and Chambers concluded that need reduction alone may serve as a reinforcer. Other

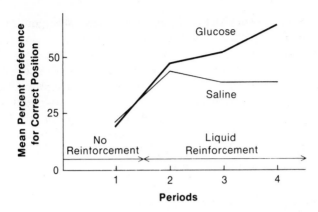

Figure 10–1. Mean percent preference for correct position during a 10-minute nonreinforcement period and three 30-minute reward periods for the glucose and saline groups.

experiments have supported this conclusion, yet have shown that food taken directly via the mouth is a far superior reinforcer compared to food directly injected into the subject's stomach (e.g., Miller & Kessen, 1952).

Evidence Against the Drive-Reduction Theory

The drive-reduction position had great appeal. Drive could be specified in terms of the need-producing operations and could be quantified in terms of the strength of motivated behavior. For example, gross locomotor activity is correlated with hunger, at least under some circumstances, and performance is generally related to the drive level.

There are, nevertheless, numerous reasons for questioning the drive-reduction position. Many studies have shown that stimuli which do not reduce biological needs, and therefore drive, may be reinforcing (see Eisenberger, 1972, for a review). For example, Sheffield and Roby (1950) found that rats could learn a T-maze response for a saccharin-solution reward. Saccharin is nonnutritive and thus could not have reduced any biological need.

Second, there are numerous experiments showing that a sensory, environmental change can reinforce behavior. Kish (1955) found that mice would press a lever in order to turn on a dim light. It is generally thought that no drive reduction is involved in this behavior, although, as discussed in Chapter 4, the effect may be quite dependent on prior sensory deprivation. In an experiment similar to Kish's, Butler (1953) showed that monkeys would perform a response in order to gain visual access to the environment outside their cage for 30 seconds. In a second study Butler (1957) showed that the effect was related to the length of prior sensory deprivation. In addition, Montgomery (1954) demonstrated that rats learned a T-maze response in order to gain access to a complex, checkerboard maze.

In summary, then, the drive-reduction theory is seriously questioned by these findings. The way in which sensory change and exploration operate to reinforce behavior is not clearly understood. A sensory deprivation state may be involved (see Fowler, 1967), which suggests that organisms may possess a need for sensory stimulation even though other biological needs, with which the drive-reduction theory dealt, are certainly not involved.

Theories Not Specifying a Reward Mechanism

As discussed previously, Thorndike postulated that "satisfying" stimuli were positive reinforcers; they increased response probability. The reinforcement mechanism, therefore, was satisfaction. As a statement concerning the mechanism of reward, however, Thorndike's theoretical Law of Effect was severely criticized, principally because satisfaction was a descriptive, circular term which could not be dealt with objectively. What does it really mean to say that a stimulus is satisfying?

In contrast to the theoretical Law of Effect the empirical Law of Effect made no such claim. The latter was a statement which simply pointed out the empirical relationship between responses and reinforcing stimuli: Responses that are followed by reinforcers increase in probability. No attempt was made to specify why, or how, the strengthening occurred. Rather, the concern was for the functional relationship between responses and stimuli.

The empirical Law of Effect is at the center of Skinner's approach to behavior. In numerous papers, he has argued that theorizing is inappropriate, and perhaps counterproductive, to the study of behavior. Instead, the more important and useful approach is an empirical one in which the functional relationships of responses to reinforcers are established. According to Skinner, a notion such as drive reduction adds very little to our understanding of behavior. Rather than spending time on such a theory, psychologists should seek to predict behavior by discovering the way in which reinforcement contingencies operate to modify it. To elaborate from Skinner's (1950) classic paper: "A science of behavior must eventually deal with behavior in its relation to certain manipulable variables. Theories, whether neural, mental, or conceptual talk about intervening steps in these relationships. But instead of prompting us to search for and explore relevant variables, they frequently have quite the opposite effect. . . . When we assert that an animal acts in a given way because it expects to receive food, then what began as the task of accounting for learned behavior becomes the task of accounting for expectancy. The problem [of accounting for expectancy] is at least equally complex and probably more difficult. We are likely to close our eyes to it and to use the theory to give us answers in place of the answers we might find through further study [p. 194]." Skinner later added ". . . The Law of Effect is no theory. It simply specifies a procedure for altering the probability of a chosen response. But when we try to say *why* reinforcement has this effect, theories arise [p. 200]."

In summary, Skinner has proposed that behavior be dealt with on an objective, descriptive level, and that questions concerning mechanisms of reward are unnecessary to an understanding of behavior. Of greater significance is an appreciation of the environmental contingencies that control behavior.

Premack's Prepotent Theory

An intriguing theory which stems from the Skinnerian tradition has been offered by Premack (1959, 1965). The theory is an empirical one and does not attempt to say why reinforcement works. Instead, it formulates a general principle, which accounts for changes in performance, based upon only a few, simple assumptions.

Premack first noted that animals perform many responses at a certain rate in an unconstrained environment. For example, given food, water, and a running wheel with no constraints upon access, animals will divide their time between eating, drinking, running, and other unrelated behaviors. The frequency or rate of each response is usually quite constant for each subject. Other situations may include different manipulanda for which individual monkeys display a fixed order of preference, although the ordering varies across monkeys (Premack, 1963b). Thus, Premack indicated that it is possible to rank different responses in terms of their relative frequency of occurrence. From that ranking, relative preference or probability of occurrence may be estimated.

Premack's theory involves several basic postulates. First, he proposed that reinforcement (as evidenced by an increase in the rate or probability of a response) always involves the contingency of one response upon another. A given response will increase in frequency (i.e., be reinforced) if access to a more probable response is made contingent on execution of that response. In the typical situation, a hungry rat will prefer eating to lever-pressing, given no constraints on its choice. Only if the more probable response (eating) is made contingent on the less probable response (lever-pressing) will a reinforcement relationship be established, in which case lever-pressing will increase in frequency. The reinforcement value of an activity, therefore, is determined by the independent probability of that activity relative to the other available activities. The greater the disparity between the probabilities or preferences, the more reinforcing the preferable response will be. Food for a satiated rat, of course, is not as reinforcing as for a hungry rat. Here, the disparity between eating and lever-pressing is less than the disparity between eating and lever-pressing for a hungry rat. In summary, to reinforce a behavior, one must simply demand that it be performed before access to a more preferable behavior is given.

Many studies support Premack's theory. For example, Premack (1963a) investigated lever-pressing in thirsty rats. First, independent response probabilities (relative time spent performing the response on a 600-minute test) were established for running in a wheel weighted either 18 or 80-gm., and for drinking sucrose solutions of either 16, 32, or 64 percent. Normally, a subject prefers the 18-gm. to the 80-gm. wheel because it is easier to operate. Also, thirsty rats prefer water to sucrose (although the opposite is true for satiated rats) or, as in this study, a lower concentration of sucrose. The response probabilities, shown on the abscissa of Figure 10-2, reflect the relative time spent performing each response on the prior baseline tests.

In the second phase of the experiment, when Premack made each of these behaviors contingent upon a lever-press, the extent of this subsequent pressing was a positive function of the response probability established earlier. The reinforcing value of each activity, in terms of lever-pressing, was a function of the relative preference of the

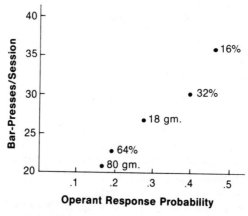

Figure 10–2. Mean lever-presses per session for each of the five reinforcing behaviors.

activity. The 18-gm. wheel activity, in which the subjects had engaged 25 percent of their time on the baseline test, reinforced lever-pressing behavior. However, a 16-percent sucrose solution, with a probability of approximately .45, was even more rewarding. In summary then, reinforcement occurred when a more probable response was made contingent on a less probable response. The extent of reinforcement reflected the disparity of the probabilities.

An important prediction which Premack's theory makes is that the typical reinforcement relationship can be reversed; in principle, wheel-running could, paradoxically, reinforce drinking if the response probabilities were reversed. This prediction was supported when Premack (1961) rationed water intake and exposure to a running wheel in order to manipulate the relative preference of each response. When the animal was satiated with water but had been deprived of access to the wheel, wheel-running reinforced drinking. That is, drinking increased when wheel-running was made contingent on drinking. Thus, Premack's experiment nicely demonstrated that there are no intrinsic reinforcers; rather, reinforcement value is a function of the relative probability of the response.

Expectancy Theory

In addition to the two theoretical approaches discussed thus far, a recent cognitive theory has been proposed by Bolles (1972). While drive-reduction theory specifies a reward mechanism in great detail, and Skinnerian approaches ignore the usefulness of discussing a reward mechanism, this new theory explicitly denies that a reinforcement mechanism underlies learning. Bolles does not claim that reinforcement, as a *procedure*, is ineffective. Rather, he denies that the change in behavior, which the reinforcement procedure produces, necessarily involves an underlying reinforcement *process*.

Failure of Reinforcement Theory

The claim that a reinforcement mechanism underlies learning is supported by the fact that reinforcement procedures work. Responses do increase in probability, suggesting that a process is involved which strengthens them. There is a growing awareness that many situations exist where typical reinforcement procedures, unaccountably, do not seem to work. Such cases strongly imply that the normal reinforcement procedure is not always sufficient for producing learning and, therefore, that an underlying mechanism of reinforcement may be absent.

One of the most notable examples where reinforcement contingencies fail concerns misbehaviors—responses which animals seem quite incapable of acquiring. Breland and Breland (1961), for example, reported that raccoons easily learned to deposit a single token in a slot to obtain food but could not perform the same response with two tokens. With continued training, the conditioned behavior deteriorated and the animals reverted to their instinctive food-getting behaviors. Apparently, reinforcement contingency was insufficient for establishing this behavior.

In contrast there are several examples in which the opposite is found: Behaviors are learned too quickly for reinforcement to have been the cause. One such example is auto-shaping, a situation in which pigeons learn to peck a disc even though no explicit shaping procedure is utilized. For example, Brown and Jenkins (1968) made it possible for pigeons to peck a disc for food. Occasionally, free food was given independent of the pigeon's behavior. No shaping procedure whatsoever was used to teach the pigeon the pecking response, yet within a short time, the subjects had acquired the response.

In another experiment, Williams and Williams (1969) arranged conditions such that pecks actually delayed the delivery of food; free food was also administered independent of the subject's behavior. Despite the fact that subjects could obtain more food by not pecking, responding continued at a high rate. Again, the reinforcement contingency was irrelevant to this behavior.

There are several examples where the reinforcement procedure is ineffective in controlling behavior. Pecking has been assumed to be a highly arbitrary response, and that fact has been cited as a virtue regarding its use in the scientific study of behavior. Yet, pecking is a very prepotent, biological response which, in some situations, is not controlled by the reinforcement contingency. Likewise, other misbehaviors occur which are not modified by reinforcement procedures. Because of these findings, Bolles has questioned the appropriateness of a reinforcement mechanism for describing the ontogeny of acquisition patterns.

Bolles claimed that these responses reflect the animal's natural behavior patterns, which occur instinctively rather than as a result of reinforcement contingencies. The misbehaviors described by Breland and Breland (1961) resembled the raccoon's natural food-getting behavior, and pecking for food topographically resembled the natural consummatory response of the pigeon. Similarly (as discussed in Chapter 5), avoidance conditioning may be based upon the animals' natural defense reactions rather than a reinforcement process (Bolles, 1970). Essentially then, Bolles' point is that in many cases animals are performing natural, innate responses—either appetitive or defensive—rather than arbitrary operants that are established or strengthened by a reinforcement process.

Expectancy

What, then, is learned by the subject in an experiment? Bolles' resolution to this problem is expectancy, a concept first used by Tolman (1932). An animal is capable of performing species-specific defensive or appetitive responses when it is brought into the experimental situation. The structure of the experiment—the events presented to the subject—produce an expectation and, in turn, the subjects' characteristic behavior. Therefore, through the presentation of contingent food (or shock), the subjects expect food (or shock) and perform their natural appetitive (or defensive) reactions. In summary, Bolles proposed that subjects learn to expect events and that they perform instinctive responses based upon those expectancies.

More specifically, Bolles suggested that two kinds of expectancies are formed. The first is an expectancy of food or some other biologically important event, like shock, as a consequence of certain events or stimuli. The contingency or correlation between these events corresponds to the CS-US correlation in classical conditioning. Bolles' theory, then, states that during classical CS-US pairings, the subject begins to expect the US. A second kind of expectancy is based on the response-reward correlation: Subjects learn to expect food or, say, shock avoidance, as a consequence to their response. This type of expectancy is based upon instrumental procedures.

The two kinds of expectancies, or stored information, correspond to the basic experimental paradigms of classical and instrumental conditioning. Although Bolles does not specify how expectancies produce performance, the subjects do perform innate, characteristic responses for the goal object based upon their expectancies. When the natural response, say, to shock, fails to produce the expected result, different expectancies are formed. For example, the subject may innately expect to reach safety by running away from shock. If running does not achieve safety, the subject will develop another response, namely freezing, i.e., the animal acquires an expectancy that freezing will achieve safety. If none of the natural-defense reactions are suitable, very poor performance will result as exemplified when subjects can avoid only by lever-pressing.

It is too soon to evaluate the impact of Bolles' theory. Clearly, a great deal of additional information will be needed. However, much evidence has already accumulated which questions the ubiquitous principle of reinforcement contingency. Moreover, the notion of expectancy is consistent with other trends in learning theory. Some researchers, for example, have cited CS-US correlation, as opposed to simple contiguity, as the mechanism of classical conditioning (Rescorla, 1967b).

Universality and the Future of Learning Theory

There are many more instances in which training procedures have proven ineffective besides the two examples cited in the preceding section. Some of these behaviors occur too quickly (as auto-shaping) and some too slowly, if at all (like misbehaviors), to be attributable to a general reinforcement process. It appears that the principles of learning described in earlier chapters fail to apply to many situations. This fact has led Seligman (1970) to propose that different classes of behavior exist and that the basic learning principles do not apply uniformly to all classes.

By basic learning principles, it is meant that arbitrary responses become conditioned by virtue of their association with reward, and arbitrary CS's become associated with US's through contiguity. Seligman suggested that stimuli and responses, in fact, are not arbitrary but are of greater or less biological significance to the animal; principles of learning, which were developed to account for arbitrary stimuli and responses, therefore, do not apply to all behaviors. An animal is more or less predisposed to learn certain responses, or CS-US relationships, through evolutionary selection. Seligman did not simply state that some responses are innate or instinctive while others are not. Rather, he indicated that the ability to learn some responses may be greater than others because those responses are more biologically significant.

Seligman proposed that there is a dimension of preparedness according to which learned behaviors can be classified. Preparedness refers to the capacity of the animal to learn specified instrumental responses or associate a CS with a US. More specifically, preparedness is defined as the amount of input (number of trials or CS-US pairings) needed to produce an output (responses) that is judged as evidence of acquisition. If a subject learns a response almost immediately with very little training, as in auto-shaping, the subject is said to be prepared to behave in that fashion. In contrast, a response that requires a great deal of training before it occurs reliably, as with misbehaviors, indicates that the subject is contraprepared to learn that response. In the middle of the preparedness dimension lies unprepared behavior, which is acquired with medium difficulty.

Seligman extended his argument by noting that the basic learning principles apply only to unprepared behaviors, and that quite different principles are required to account for prepared and contraprepared responses. To substantiate his theory, he cited several examples which illustrate how inadequately the present basic learning principles account for prepared and contraprepared behaviors (see Seligman &

Hager, 1972, for additional examples and discussion). One such example is acquired taste aversion. As discussed in Chapter 2, subjects are able to associate flavored water with the experience of nausea even though several hours separate those two events. The CS-US interval is so long that it is difficult to view this learning within the context of the traditional classical conditioning literature. Stated another way, taste aversion is acquired so quickly, despite long delays in the US presentation, that the principle of repeated CS-US pairings seems inappropriate for explaining the effect. According to Seligman, acquired taste aversion is an example of prepared learning. The organism is predisposed to associate illness with relevant, food-related stimuli because those kinds of associations are of extreme biological significance to the animal's survival; natural selection favors animals that respond in this self-preserving manner.

In contrast to such a prepared response, animals are virtually unable to associate typical exteroceptive stimuli, such as lights and tones, with subsequent illness (e.g., Garcia & Koelling, 1966). The animals are contraprepared to learn that association because it is biologically less significant insofar as animals do not normally experience such artificial lights and tones in their natural habitat.

The fact that preparedness is related to the organism's evolutionary history was elegantly shown in a study by Wilcoxon, Dragoin, and Kral (1971). Rats acquired aversion to flavored, but not colored, water, both of which were followed by illness. In other words, taste was more important than a visual cue for the rats. In contrast, quail, for which the relationship between visual cues and food is of great biological significance, learned to avoid the colored water even more than the flavored water. For quail, the visual cue was more potent than the taste cue, reflecting the relative importance of vision. In summary, the ability to associate cues varies not only with the types of stimuli used (differing biological significance), but also with species (which differ in their evolutionary history and, therefore, their learning capacities).

The preparedness dimension can also be applied to instrumental behavior. The examples of auto-shaping and misbehaviors cited previously illustrate prepared and contraprepared responses, and others could be added to the list. For example, rats can learn to passively avoid a shock in one trial by remaining on a platform (prepared response), yet have considerable difficulty in learning to avoid shock by lever-pressing (contraprepared response; D'Amato & Schiff, 1964). Similarly, pigeons readily learn to peck a disc for food but require considerable training before they will peck to avoid shock (e.g., Hoffman & Fleshler, 1959; Azrin, 1959).

In limiting the current principles of learning to the unprepared class of behaviors, Seligman has not invalidated learning theory or emasculated learning research. On the contrary, his thesis opens up new possibilities for theory and additional avenues of investigation. Added principles, which account for prepared and contraprepared learning, will probably enrich our understanding of learned behavior. Although ethologists have for years emphasized the importance of evolutionary history for species-specific behavior, many learning psychologists are only now beginning to recognize the implications of evolution as it relates to the biological boundaries of learned behaviors and a theory of learning in general.

Utility of Learning Theory

While not true of all areas of learning research, the basic Skinnerian approach has been moderately successful regarding the application of learning principles and techniques to some societal issues. These issues, of course, involve behavioral change, which is efficiently managed through the appropriate arrangement of reinforcement schedules. The principles of learning which have been shown to modify animal performance have, in turn, been applied to human performance. Efficient or knowledgeable manipulation of these principles has facilitated human behavioral change in a variety of important ways.

Teaching Machines

One of the most notable contributions stemming from the Skinnerian tradition has been the teaching machine, or more generally, programmed learning (e.g., Skinner, 1958). A body of material is broken down into small, discrete units, each of which builds from the previous units and identifies a single point to be mastered. Usually, the unit is in the form of a question or statement in which a key word is missing. The student fills in the blank and then consults a separate list to determine if the answer was correct. If the answer was incorrect, the student is often instructed to return to a previous point and begin again. Otherwise, the student continues to master subsequent units.

The important feature of programmed learning is that it maximizes many principles of learning. Progress is gradual and the student proceeds to new material only after mastering prior material. The units, or statements, themselves are relatively easy, thus increasing the probability of success and minimizing the frustration and anxiety of failure. Quite clearly, success is reinforcing to the student and to that extent, reinforcement magnitude is maximized while delay of reward is minimized. Furthermore, the units, and their sequential development, identify for the student the important concepts which must be mastered, thus eliminating competing and inappropriate responses. In general, programmed learning establishes a more efficient and productive context in which material can be mastered. Presumably, any body of material can be programed, provided the concepts and answers can be identified.

The teaching machine is simply an automated apparatus for presenting programmed material. Typically, a unit appears in a window along with several alternative answers. The student chooses the appropriate answer by pressing a button corresponding to his choice, and the machine advances to reveal the correct answer and the next unit. These machines have been used in many ways, from teaching small children the alphabet to instructing college students in foreign languages. More recently, computers have been used to facilitate both the presentation of material and the rapid feedback following an answer (see Atkinson, 1968).

Programmed instruction has been used successfully in many instances, although some types of materials are more conducive to programming than others. One principal advantage seems to be the reduced time required for a student to master a given body of material, as compared to the time required using more conventional techniques. However, there is some question as to whether programmed learning leads to better retention of the material (see Nash, Muczyk, & Vettori, 1971, for a review).

An important variation of the programmed learning approach was first developed by Keller (1968). Keller devised a plan in which an entire course—for example, Introductory Psychology—could be programed. The course was divided into a number of units, each of which covered a discrete body of material. The units were small enough so that, by correctly answering questions on a study guide which identified the important concepts, the material could be easily mastered. Once the student felt he was prepared, he would volunteer for a quiz. If he passed, he continued on to the next unit. If not, the proctor, who graded the quiz, diagnosed his weaknesses and the student would try again after reviewing the material. No penalty was levied for failing the quiz; instead, the subject was reinforced by eventually passing and continuing to the next unit.

The Keller Plan, as it is now called, appears to be quite successful and several studies indicate that students both learn more and enjoy the approach (e.g., McMichael & Corey, 1969). There are several reasons for such success: First, students are active participants rather than passive listeners in a lecture hall; second, they may proceed at their own pace without the fear of failing or falling behind; third, they proceed to new material only after mastering prior material; and finally, students are given continuous feedback on strengths and weaknesses. Although the Keller Plan deserves further investigation, it is an innovative and highly successful educational trend based upon the more general concepts of programmed learning. At the very least, programmed learning approaches have challenged learning theorists to specify in detail the factors in a student's environment which promote efficient performance.

Behavior Therapy

Just as programmed learning represents the application of learning principles to one segment of human behavior, behavior therapy applies these principles to abnormal behavior. The use of behavior therapy, or behavior modification techniques, is exceedingly widespread (see Franks, 1969; Krasner & Ullmann, 1965; Lazarus, 1971; London, 1972; Wenrich, 1970; and Wolpe, 1958, for discussions of behavior therapy and its implications).

According to these therapists, abnormal behavior follows the same basic principles of learning as other behaviors do. It is incumbent upon the therapist, then, to isolate the maladaptive behavior and to treat it through, say, counterconditioning and extinction. It is often claimed that maladaptive behaviors, such as so-called neuroses, are based upon fear. Quite clearly, an understanding of the factors that create, and eliminate, fear would aid in the treatment of fear-motivated behavior.

In actual practice, maladaptive behaviors have been treated in a variety of ways. The differences in approach are often subtle, yet each generally represents an attempt to extinguish the neurotic behavior while reinforcing an adaptive response. It should be emphasized that these are treatments of behavior *per se,* and no appeal whatsoever is made to internal events or states such as "will," "id," "superego," or other hypothetical constructs which do not lend themselves to objective analysis.

One of the major techniques is implosive therapy (Stampfl & Levis, 1967). Essentially, the patient attempts to imagine or visualize the fear-provoking cues at the suggestion of the therapist. The resulting anxiety is then extinguished since no punishing consequences (i.e., the US) are experienced. This technique stems from the work on response prevention, or flooding, discussed in Chapter 7 (Baum, 1969). The idea behind this therapy is that the maladaptive behavior will disappear if the underlying fear is extinguished.

A similar, and older, technique is termed systematic desensitization, which involves not only the extinction of anxiety, but also the counterconditioning of fear cues to responses associated with relaxation. The patient is first trained to relax the muscles in a systematic fashion. Once this is achieved, fear cues are imagined or visualized and juxtaposed with the relaxed state, and the effect is extinction of fear and a gradual acquisition of relaxation responses to the fear cues.

For example, Kushner (1965) reported about a man who developed a phobia about cars following an accident. He was anxious, and, ultimately, unable to drive. In the therapy session, the patient was taught to relax deeply; then, he was asked to visualize a car. When he was able to maintain his relaxed state, other imaginal stimuli were presented until, after several sessions, the patient could visualize driving (he could even visualize a situation similar to his previous accident) with little discomfort. In several weeks' time, the patient had resumed a normal pattern of behavior and could drive his car without feeling anxious. In summary, the patient had extinguished his anxiety and had learned to relax in the presence of previous fear cues.

Although these behavior therapy techniques have been used more widely in treating phobias and other neurotic disorders, some success has been found using these methods in treating psychotic patients. For example, a classic study by Ayllon and Haughton (1962) dealt with schizophrenics. Most of these patients showed some refusal to eat; many were spoon-fed or given intravenous feeding by nurses. The authors argued that such sympathetic treatment may have actually encouraged their behavior of not eating. Therefore, the first part of the treatment consisted of discontinuing the spoon-feeding and reinforcing eating behavior with food. Patients were required to enter the dining room during a given 30-minute period or else go without food for that meal; this time was reduced in subsequent weeks to 20, 15, or 5 minutes. The results showed that the percentage of meals eaten increased considerably for those patients with eating problems, suggesting that food had been an effective reinforcer.

In subsequent experiments, Ayllon and Haughton extended this work by showing that patients could learn to cooperate with other patients if the reinforcers were structured properly. The patients had to obtain a penny from the nurse in order to be admitted to the dining room. To get the penny, two of them had to push buttons simultaneously which lit a light and sounded a buzzer. Not only did cooperation develop but also verbal exchanges with nurses or other patients increased substantially. These results therefore indicate that by applying basic principles of learning, adaptive responses may be taught to patients who normally do not display such behavior. While the Ayllon and Haughton studies are particularly dramatic, these techniques have also been used successfully in more normal settings, as in modifying or extinguishing smoking behavior.

Behavior therapy has become a powerful tool in the control of certain disorders. Although a great deal of information still remains to be uncovered about effective therapy, the success of this approach is well documented.

Summary

Hull attempted to specify a mechanism for learning by claiming that reinforcement resulted from a reduction in drive. His theory has not been supported by many experiments which show that sensory and other nonbiological stimuli (which do not reduce a need) can reinforce behavior.

In contrast to Hull, Skinner's general approach does not specify a mechanism of reward. Rather, he stresses the importance of discovering the functional relationship between stimuli and responses without using theoretical constructs. An important theory by Premack, which stems from the Skinnerian approach, also does not specify a reward mechanism, but does establish an empirical principle which accounts for performance changes. A response will increase in frequency if access to a more probable response is made contingent on its execution. An important demonstration of this principle involved a reversal of a normal contingency; in this case, wheel-running reinforced drinking.

A third approach has been proposed by Bolles. He denies that a reinforcement process necessarily underlies all behavior. One impetus to his theory is that there are responses which seem not to be affected by reward contingency. In light of these findings, Bolles suggested that subjects learn expectancies which alter their species-specific behavior.

There are many behaviors which are conditioned too quickly or too slowly to be considered examples of typical classical or instrumental conditioning. This fact led Seligman to propose that different classes of behavior exist. Prepared behavior is innate-like; contraprepared behavior is learned with great difficulty; unprepared behavior, to which contemporary learning principles apply, is learned with moderate difficulty. Current research confirms Seligman's position insofar as learning abilities vary according to the specific response and its relationship to the subject's natural behavior.

Many principles of learning have successfully been applied to societal issues. One example is the development of teaching machines or programed learning. Another example is behavior therapy by means of which maladaptive behaviors are extinguished while adaptive responses are reinforced. Various techniques for achieving these ends have been successfully applied to phobias as well as to some psychotic behaviors.

References

References

Abbott, D. W., & Price, L. E. Stimulus generalization of the conditioned eyelid response to structurally similar nonsense syllables. *Journal of Experimental Psychology,* 1964, *68,* 368–371.

Adelman, H. M., & Maatsch, J. L. Resistance to extinction as a function of the type of response elicited by frustration. *Journal of Experimental Psychology,* 1955, *50,* 61–65.

Amsel, A. The role of frustrative nonreward in noncontinuous reward situations. *Psychological Bulletin,* 1958, *55,* 102–119.

Amsel, A. Frustrative nonreward in partial reinforcement and discrimination learning: Some recent history and a theoretical extension. *Psychological Review,* 1962, *69,* 306–328.

Amsel, A. Behavioral habituation, counterconditioning, and a general theory of persistence. In A. H. Black and W. F. Prokasy (Eds.) *Classical conditioning II: Current research and theory.* New York: Appleton-Century-Crofts, 1972.

Amsel, A., Hug, J. J., & Surridge, C. T. Number of food pellets, goal approaches, and the partial reinforcement effect after minimal acquisition. *Journal of Experimental Psychology,* 1968, *77,* 530–534.

Amsel, A., & Roussel, J. Motivational properties of frustration: I. Effect on a running response of the addition of frustration to the motivational complex. *Journal of Experimental Psychology,* 1952, *43,* 363–368.

Anger, D. The role of temporal discrimination in the reinforcement of Sidman avoidance behavior. *Journal of the Experimental Analysis of Behavior,* 1963, *6,* 477–506.

Annau, Z., & Kamin, L. J. The conditioned emotional response as a function of intensity of the US. *Journal of Comparative and Physiological Psychology,* 1961, *54,* 428–432.

Armus, H. L., & Garlich, M. M. Secondary reinforcement strength as a function of schedule of primary reinforcement. *Journal of Comparative and Physiological Psychology,* 1961, *54,* 56–58.

Atkinson, R. C. Computerized instruction and the learning process. *American Psychologist,* 1968, *23,* 225–239.

Ayllon, T., & Haughton, E. Control of the behavior of schizophrenics by food. *Journal of the Experimental Analysis of Behavior,* 1962, *5,* 343–352.

Ayres, J. J. B. Conditioned suppression and the information hypothesis. *Journal of Comparative and Physiological Psychology,* 1966, *62,* 21–25.

Azrin, N. H. Some effects of two intermittent schedules of immediate and non-immediate punishment. *Journal of Psychology,* 1956, *42,* 3–21.

Azrin, N. H. Some notes on punishment and avoidance. *Journal of the Experimental Analysis of Behavior,* 1959, *2,* 260.

Azrin, N. H., & Holz, W. C. Punishment during fixed-interval reinforcement. *Journal of the Experimental Analysis of Behavior,* 1961, *4,* 343–347.

Barnes, G. W., & Kish, G. B. Reinforcing properties of the onset of auditory stimulation. *Journal of Experimental Psychology,* 1961, *62,* 164–170.

Baron, A. Delayed punishment of a runway response. *Journal of Comparative and Physiological Psychology,* 1965, *60,* 131–134.

Barry, H. Effects of strength of drive on learning and on extinction. *Journal of Experimental Psychology,* 1958, *55,* 473–481.

Barry, H. Effects of drive strength on extinction and spontaneous recovery. *Journal of Experimental Psychology,* 1967, *73,* 419–421.

Baum, M. Efficacy of response prevention (flooding) in facilitating the extinction of an avoidance response in rats: The effect of overtraining the response. *Behavior Research and Therapy,* 1968, *6,* 197–203.

Baum, M. Extinction of an avoidance response following response prevention: Some parametric investigations. *Canadian Journal of Psychology,* 1969, *23,* 1–10.

Baum, M. Extinction of avoidance responding through response prevention (flooding). *Psychological Bulletin,* 1970, *74,* 276–284.

Beck, R. C. On secondary reinforcement and shock termination. *Psychological Bulletin,* 1961, *58,* 24–45.

Beck, S. B. Eyelid conditioning as a function of CS intensity, UCS intensity, and manifest anxiety scale score. *Journal of Experimental Psychology,* 1963, *66,* 429–438.

Beecroft, R. S. *Classical conditioning.* Goleta, Calif.: Psychonomic Press, 1966.

Behrend, E. R., & Bitterman, M. E. Avoidance conditioning in the goldfish: Exploratory studies of the CS-US interval. *American Journal of Psychology,* 1962, *75,* 18–34.

Bekhterev, V. M. *La Psychologie objective.* Paris: Alcan, 1913.

Benedict, J. O., & Ayres, J. J. B. Factors affecting conditioning in the truly random control procedure in the rat. *Journal of Comparative and Physiological Psychology,* 1972, *78,* 323–330.

Bersh, P. J. The influence of two variables upon the establishment of a secondary reinforcer for operant responses. *Journal of Experimental Psychology,* 1951, *41,* 62–73.

Bintz, J. Time-dependent memory deficits of aversively motivated behavior. *Learning and Motivation,* 1970, *1,* 382–390.

Bitterman, M. E. The CS-US interval in classical and avoidance conditioning. In W. F. Prokasy (Ed.) *Classical conditioning: A symposium.* New York: Appleton-Century-Crofts, 1965.

Bitterman, M. E., Fedderson, W. E., & Tyler, D. W. Secondary reinforcement and the discrimination hypothesis. *American Journal of Psychology,* 1953, *66,* 456–464.

Bitterman, M. E., & Schoel, W. M. Instrumental learning in animals: Parameters of reinforcement. *Annual Review of Psychology,* 1970, *21,* 367–436.

Black, A. H. Heart rate changes during avoidance learning in dogs. *Canadian Journal of Psychology,* 1959, *13,* 229–242.

Black, A. H. The effect of CS-US interval on avoidance conditioning in the rat. *Canadian Journal of Psychology,* 1963, *17,* 174–182.

Black, A. H., Carlson, N. J., & Solomon, R. L. Exploratory studies of the conditioning of autonomic responses in curarized dogs. *Psychological Monographs,* 1962, *76* (Whole No. 548).

Black, A. H., & DeToledo, L. The relationship among classically conditioned responses: Heart rate and skeletal behavior. In A. H. Black and W. F. Prokasy (Eds.), *Classical conditioning II: Current research and theory.* New York: Appleton-Century-Crofts, 1972.

Black, R. W. Shifts in magnitude of reward and contrast effects in instrumental and selective learning. *Psychological Review*, 1968, *75*, 114–126.

Blodgett, H. C. The effect of the introduction of reward upon the maze performance of rats. *University of California Publications of Psychology*, 1929, *4*, 113–134.

Bloomfield, T. M. Behavioral contrast and the peak shift. In R. M. Gilbert and N. S. Sutherland (Eds.), *Animal discrimination learning.* New York: Academic Press, 1969.

Bolles, R. C. *Theory of motivation.* New York: Harper & Row, 1967.

Bolles, R. C. Species-specific defense reactions and avoidance learning. *Psychological Review*, 1970, *77*, 32–48.

Bolles, R. C. Reinforcement, expectancy, and learning. *Psychological Review*, 1972, *79*, 394–409.

Bolles, R. C., & Grossen, N. E. Effects of an informational stimulus on the acquisition of avoidance behavior in rats. *Journal of Comparative and Physiological Psychology*, 1969, *68*, 90–99.

Bolles, R. C., Grossen, N. E., Hargrave, G. E., & Duncan, P. M. Effects of conditioned appetitive stimuli on the acquisition and extinction of a runway response. *Journal of Experimental Psychology*, 1970, *85*, 138–140.

Bolles, R. C., Stokes, L. W., & Younger, M. S. Does CS termination reinforce avoidance behavior? *Journal of Comparative and Physiological Psychology*, 1966, *62*, 201–207.

Bolles, R. C., & Warren, J. A. The acquisition of bar press avoidance as a function of shock intensity. *Psychonomic Science*, 1965, *3*, 297–298.

Bolles, R. C., Warren, J. A., & Ostrov, N. The role of the CS-US interval in bar press avoidance learning. *Psychonomic Science*, 1966, *6*, 113–114.

Boren, J. J., Sidman, M., & Herrnstein, R. J. Avoidance, escape, and extinction as functions of shock intensity. *Journal of Comparative and Physiological Psychology*, 1959, *52*, 420–425.

Boring, F. G. *A history of experimental psychology.* New York: Appleton-Century-Crofts, 1957.

Boroczi, G., Storms, L. H., & Broen, W. E. Response suppression and recovery of responding at different deprivation levels as a function of intensity and duration of punishment. *Journal of Comparative and Physiological Psychology*, 1964, *58*, 456–459.

Bower, G. H. A contrast effect in differential conditioning. *Journal of Experimental Psychology*, 1961, *62*, 196–199.

Bower, G. H., Fowler, H., & Trapold, M. A. Escape learning as a function of amount of shock reduction. *Journal of Experimental Psychology*, 1959, *58*, 482–484.

Bower, G. H., Starr, R., & Lazarovitz, L. Amount of response-produced change in the CS and avoidance learning. *Journal of Comparative and Physiological Psychology*, 1965, *59*, 13–17.

Breland, K., & Breland, M. The misbehavior of animals. *American Psychologist*, 1961, *16*, 681–684.

Brogden, W. J., Lipman, E. A., & Culler, E. The role of incentive in conditioning and extinction. *American Journal of Psychology*, 1938, *51*, 100–117.

Brown, J. L. The effect of drive on learning with secondary reinforcement. *Journal of Comparative and Physiological Psychology*, 1956, *49*, 254–260.

Brown, J. S. Factors affecting self-punitive locomotor behavior. In B. A. Campbell and R. M. Church (Eds.), *Punishment and aversive behavior.* New York: Appleton-Century-Crofts, 1969.

Brown, J. S. Self-punitive behavior with a distinctly marked punishment zone. *Psychonomic Science*, 1970, *21*, 161–163.

Brown, J. S., Martin, R. C., & Morrow, M. W. Self-punitive behavior in the rat: Facilitative effects of punishment on resistance to extinction. *Journal of Comparative and Physiological Psychology*, 1964, *57*, 127–133.

Brown, P. L., & Jenkins, H. M. Auto-shaping of the pigeon's key-peck. *Journal of the Experimental Analysis of Behavior*, 1968, *11*, 1–8.

Brush, F. R. The effects of shock intensity on the acquisition and extinction of an avoidance response in dogs. *Journal of Comparative and Physiological Psychology*, 1957, *50*, 547–552.

Brush, F. R. The effects of intertrial interval on avoidance learning in the rat. *Journal of Comparative and Physiological Psychology*, 1962, *55*, 888–892.

Brush, F. R. Avoidance learning as a function of time after fear conditioning and unsignaled shock. *Psychonomic Science*, 1964, *1*, 405–406.

Brush, F. R. The effects of amount of signaled escape training on subsequent avoidance learning. *Psychonomic Science*, 1970, *21*, 51–52.

Brush, F. R. Retention of aversively motivated behavior. In F. R. Brush (Ed.), *Aversive conditioning and learning.* New York: Academic Press, 1971.

Brush, F. R., Brush, E. S., & Solomon, R. L. Traumatic avoidance learning: The effects of the CS-US interval with a delayed-conditioning procedure. *Journal of Comparative and Physiological Psychology*, 1955, *48*, 285–293.

Brush, F. R., Meyer, J. S., & Palmer, M. E. Effects of kind of prior training and intersession interval upon subsequent avoidance learning. *Journal of Comparative and Physiological Psychology*, 1963, *56*, 539–545.

Bull, J. A. An interaction between appetitive Pavlovian CSs and instrumental avoidance responding. *Learning and Motivation*, 1970, *1*, 18–26.

Butler, R. A. Discrimination learning by rhesus monkeys to visual exploration motivation. *Journal of Comparative and Physiological Psychology*, 1953, *46*, 95–98.

Butler, R. A. The effect of deprivation of visual incentives on visual exploration motivation in monkeys. *Journal of Comparative and Physiological Psychology*, 1957, *50*, 177–179.

Butter, C. M., & Campbell, B. A. Running speed as a function of successive reversals in hunger drive level. *Journal of Comparative and Physiological Psychology*, 1960, *53*, 52–54.

Butter, C. M., & Thomas, D. R. Secondary reinforcement as a function of the amount of primary reinforcement. *Journal of Comparative and Physiologi-*

cal Psychology, 1958, *51,* 346–348.

Camp, D. S., Raymond, G. A., & Church, R. M. Temporal relationship between response and punishment. *Journal of Experimental Psychology,* 1967, *74,* 114–123.

Campbell, B. A. Developmental studies of learning and motivation in infra-primate mammals. In H. W. Stevenson, E. H. Hess, and H. L. Rheingold (Eds.), *Early behavior: Comparative and developmental approaches.* New York: Wiley, 1967.

Campbell, B. A., & Kraeling, D. Response strength as a function of drive level and amount of drive reduction. *Journal of Experimental Psychology,* 1953, *45,* 97–101.

Campbell, B. A., & Kraeling, D. Response strength as a function of drive level during training and extinction. *Journal of Comparative and Physiological Psychology,* 1954, *47,* 101–103.

Campbell, B. A., Smith, N. F., & Misanin, J. R. Effects of punishment on extinction of avoidance behavior: Avoidance-avoidance conflict or vicious circle behavior. *Journal of Comparative and Physiological Psychology,* 1966, *62,* 495–498.

Campbell, B. A., & Spear, N. E. Ontogeny of memory. *Psychological Review,* 1972, *79,* 215–236.

Capaldi, E. D. Simultaneous shifts in reward magnitude and level of food deprivation. *Psychonomic Science,* 1971, *23,* 357–359.

Capaldi, E. J. Effect on N-length, number of different N-lengths, and number of reinforcements on resistance to extinction. *Journal of Experimental Psychology,* 1964, *68,* 230–239.

Capaldi, E. J. Partial reinforcement: A hypothesis of sequential effects. *Psychological Review,* 1966, *73,* 459–477.

Capaldi, E. J. A sequential hypothesis of instrumental learning. In K. W. Spence and J. T. Spence (Eds.), *The psychology of learning and motivation,* Vol. 1. New York: Academic Press, 1967.

Capaldi, E. J., & Deutsch, E. A. Effects of severely limited acquisition training and pretraining on the partial reinforcement effect. *Psychonomic Science,* 1967, *9,* 171–172.

Capaldi, E. J., & Hart, D. Influence of a small number of partial reinforcement training trials on resistance to extinction. *Journal of Experimental Psychology,* 1962, *64,* 166–171.

Capehart, J., Viney, W., & Hulicka, I. M. The effect of effort upon extinction. *Journal of Comparative and Physiological Psychology,* 1958, *51,* 505–507.

Caul, W. F., Miller, R. E., & Banks, J. H. Effect of US intensity on heart rate in delay conditioning and pseudoconditioning. *Psychonomic Science,* 1970, *19,* 15–17.

Champion, R. A. Stimulus-intensity effects in response evocation. *Psychological Review,* 1962, *69,* 428–449.

Church, R. M. The varied effects of punishment on behavior. *Psychological Review,* 1963, *70,* 369–402.

Church, R. M. Response suppression. In B. A. Campbell and R. M. Church (Eds.), *Punishment and aversive behavior.* New York: Appleton-Century-Crofts, 1969.

Coate, W. B. Effect of deprivation on postdiscrimination stimulus generalization in the rat. *Journal of Comparative and Physiological Psychology,* 1964, *57,* 134–138.

Cohen, P. S. Interference effects of escapable shock upon subsequent acquisition of escape-avoidance responding. *Journal of Comparative and Physiological Psychology,* 1970, *71,* 484–486.

Coppock, H. W., & Chambers, R. M. Reinforcement of position preference by automatic intravenous injections of glucose. *Journal of Comparative and Physiological Psychology,* 1954, *47,* 355–357.

Coppock, W. J. Pre-extinction in sensory preconditioning. *Journal of Experimental Psychology,* 1958, *55,* 213–219.

Coughlin, R. C. The aversive properties of withdrawing positive reinforcement: A review of the recent literature. *Psychological Record,* 1972, *22,* 333–354.

Coulter, X., Riccio, D. C., & Page, H. A. Effects of blocking an instrumental avoidance response: Facilitated extinction but persistence of fear. *Journal of Comparative and Physiological Psychology,* 1969, *68,* 377–381.

Cousins, L. S., Zamble, E., Tait, R. W., & Suboski, M. D. Sensory preconditioning in curarized rats. *Journal of Comparative and Physiological Psychology,* 1971, *77,* 152–154.

Cravens, R. W., & Renner, K. E. Conditioned appetitive drive states: Empirical evidence and theoretical status. *Psychological Bulletin,* 1970, *73,* 212–220.

Crespi, L. P. Quantitative variations of incentive and performance in the white rat. *American Journal of Psychology,* 1942, *55,* 467–517.

Crider, A. B., Schwartz, G., & Shnidman, S. On the criteria for instrumental autonomic conditioning: A reply to Katkin and Murray. *Psychological Bulletin,* 1969, *71,* 455–461.

Crowder, W. F., Gill, K., Hodge, C. C., & Nash, F. A. Secondary reinforcement or response facilitation? II. Response acquisition. *Journal of Psychology,* 1959, *48,* 303–306.

Crum, J., Brown, W. L., & Bitterman, M. E. The effect of partial and delayed reinforcement on resistance to extinction. *American Journal of Psychology,* 1951, *64,* 228–237.

D'Amato, M. R. Secondary reinforcement and magnitude of primary reinforcement. *Journal of Comparative and Physiological Psychology,* 1955, *48,* 378–380.

D'Amato, M. R., & Fazzaro, J. Discriminated lever press avoidance learning as a function of type and intensity of shock. *Journal of Comparative and Physiological Psychology,* 1966, *61,* 313–315.

D'Amato, M. R., Fazzaro, J., & Etkin, M. Discriminated bar-press avoidance maintenance and extinction in rats as a function of shock intensity. *Journal of Comparative and Physiological Psychology,* 1967, *63,* 351–354.

D'Amato, M. R., Fazzaro, J., & Etkin, M. Anticipatory responding and avoidance discrimination as factors in avoidance conditioning. *Journal of Experimental Psychology,* 1968, *77,* 41–47.

D'Amato, M. R., Lachman, R., & Kivy, P. Secondary reinforcement as affected by reward schedule and the testing situation. *Journal of Comparative and Physiological Psychology,* 1958, *51,* 737–741.

D'Amato, M. R., & Schiff, D. Long-term discriminated avoidance performance in the rat. *Journal of Comparative and Physiological Psychology,* 1964, *57,* 123–126.

Das, J. P. *Verbal conditioning and behavior.* London: Pergamon Press, 1969.

Davis, H., & Kreuter, C. Conditioned suppression of an avoidance response by a stimulus paired with food. *Journal of the Experimental Analysis of Behavior,* 1972, *17,* 277–285.

Deane, G. E. Cardiac conditioning in the albino rabbit using three CS-UCS intervals. *Psychonomic Science,* 1965, *3,* 119–120.

Deese, J. The extinction of a discrimination without performance of the choice response. *Journal of Comparative and Physiological Psychology,* 1951, *44,* 362–366.

Delude, L. A. The vicious circle phenomenon: A result of measurement artifact. *Journal of Comparative and Physiological Psychology,* 1969, *69,* 246–252.

Denny, M. R. The "Kamin effect" in avoidance conditioning. *American Psychologist,* 1958, *13,* 419. (Abstract)

Denny, M. R., & Ditchman, R. E. The locus of maximal "Kamin effect" in rats. *Journal of Comparative and Physiological Psychology,* 1962, *55,* 1069–1070.

Desiderato, O., Butler, B., & Meyer, C. Changes in fear generalization gradients as a function of delayed testing. *Journal of Experimental Psychology,* 1966, *72,* 678–682.

Desiderato, O., & Wassarman, M. E. Incubation of anxiety: Effects on generalization gradients. *Journal of Experimental Psychology,* 1967, *74,* 506–510.

Dews, P. B. The effect of multiple S_Δ periods on responding on a fixed-interval schedule. *Journal of the Experimental Analysis of Behavior,* 1962, *5,* 369–374.

DiCara, L. V., & Miller, N. E. Instrumental learning of vasomotor responses by rats: Learning to respond differentially in the two ears. *Science,* 1968, *159,* 1485–1486. (a)

DiCara, L. V., & Miller, N. E. Changes in heart rate instrumentally learned by curarized rats as avoidance responses. *Journal of Comparative and Physiological Psychology,* 1968, *65,* 8–12. (b)

DiLollo, V. Runway performance in relation to runway-goal-box similarity and changes in incentive amount. *Journal of Comparative and Physiological Psychology,* 1964, *58,* 327–329.

Dinsmoor, J. A. A quantitative comparison of the discriminative and reinforcing functions of a stimulus. *Journal of Experimental Psychology,* 1950, *40,* 458–472.

Dinsmoor, J. A. Punishment: I. The avoidance hypothesis. *Psychological Review,* 1954, *61,* 34–46.

Dinsmoor, J. A. Punishment: II. An interpretation of empirical findings. *Psychological Review,* 1955, *62,* 96–105.

Dinsmoor, J. A., & Clayton, M. H. Chaining and secondary reinforcement based on escape from shock. *Journal of the Experimental Analysis of Behavior,* 1963, *6,* 75–80.

Doty, R. W., & Giurgea, C. Conditioned reflexes established by coupling electrical excitation of two cortical areas. In J. F. Delafresnaye (Ed.), *Brain mechanisms and learning.* Oxford: Blackwell Scientific Publications, 1961.

Egger, M. D., & Miller, N. E. Secondary reinforcement in rats as a function of information value and reliability of the stimulus. *Journal of Experimental Psychology,* 1962, *64,* 97–104.

Egger, M. D., & Miller, N. E. When is a reward reinforcing?: An experimental study of the information hypothesis. *Journal of Comparative and Physiological Psychology,* 1963, *56,* 132–137.

Ehrenfreund, D. A study of the transposition gradient. *Journal of Experimental Psychology,* 1952, *43,* 81–87.

Eisenberger, R. Explanation of rewards that do not reduce tissue needs. *Psychological Bulletin,* 1972, *77,* 319–339.

Engel, B. T. Operant conditioning of cardiac function: A status report. *Psychophysiology,* 1972, *9,* 161–177.

Estes, W. K. An experimental study of punishment. *Psychological Monographs,* 1944, *57* (Whole No. 263).

Estes, W. K., & Skinner, B. F. Some quantitative properties of anxiety. *Journal of Experimental Psychology,* 1941, *29,* 390–400.

Farber, I. E. Response fixation under anxiety and nonanxiety conditions. *Journal of Experimental Psychology,* 1948, *38,*111–131.

Feirstein, A. R., & Miller, N. E. Learning to resist pain and fear: Effects of electric shock before versus after reaching goal. *Journal of Comparative and Physiological Psychology,* 1963, *56,* 797–800.

Felton, M., & Lyon, D. O. The postreinforcement pause. *Journal of the Experimental Analysis of Behavior,* 1966, *9,* 131–134.

Ferster, C. B. Sustained behavior under delayed reinforcement. *Journal of Experimental Psychology,* 1953, *45,* 218–224.

Ferster, C. B., & Appel, J. B. Punishment of S_Δ responding in matching to sample by time out from positive reinforcement. *Journal of the Experimental Analysis of Behavior,* 1961, *4,* 45–56.

Ferster, C. B., & Skinner, B. F. *Schedules of reinforcement.* New York: Appleton-Century-Crofts, 1957.

Fitzgerald, R. D., & Teyler, T. J. Trace and delayed heart-rate conditioning in rats as a function of US intensity. *Journal of Comparative and Physiological Psychology,* 1970, *70,* 242–253.

Fowler, H. Satiation and curiosity. In K. W. Spence and J. T. Spence (Eds.), *The psychology of learning and motivation,* Vol. 1. New York: Academic Press, 1967.

Fowler, H., Spelt, P. F., & Wischner, G. J. Discrimination performance as affected by training procedure, problem difficulty, and shock for the correct response. *Journal of Experimental Psychology,* 1967, *75,* 432–436.

Fowler, H., & Trapold, M. A. Escape performance as a function of delay of reinforcement. *Journal of Experimental Psychology,* 1962, *63,* 464–467.

Fowler, H., & Wischner, G. J. Discrimination performance as affected by problem difficulty and shock for either the correct or incorrect response. *Journal of Experimental Psychology,* 1965, *69,* 413–418.

Fowler, H., & Wischner, G. J. The varied functions of punishment in discrimination learning. In B. A. Campbell and R. M. Church, *Punishment and aversive behavior.* New York: Appleton-Century-Crofts, 1969.

Fox, R. E., & King, R. A. The effects of reinforcement scheduling on the strength of a secondary reinforcer. *Journal of Comparative and Physiological Psychology,* 1961, *54,* 266–269.

Franchina, J. J. Escape behavior and shock intensity: Within-subject versus between-groups comparisons. *Journal of Comparative and Physiological Psychology,* 1969, *69,* 241–245.

Franks, C. M. *Behavior therapy: Appraisal and status.* New York: McGraw-Hill, 1969.

Friedman, H., & Guttman, N. Further analysis of the various effects of discrimination training upon stimulus generalization gradients. In D. I. Mostofsky (Ed.), *Stimulus generalization.* Stanford, Calif.: Stanford University Press, 1965.

Ganz, L., & Riesen, A. H. Stimulus generalization to hue in the dark-reared macaque. *Journal of Comparative and Physiological Psychology,* 1962, *55,* 92–99.

Garcia, J., & Koelling, R. Relation of cue to consequence in avoidance learning. *Psychonomic Science,* 1966, *4,* 123–124.

Garcia, J., McGowan, B. K., & Green, K. F. Biological constraints on conditioning. In A. H. Black and W. F. Prokasy (Eds.), *Classical conditioning II: Current research and theory.* New York: Appleton-Century-Crofts, 1972.

Gavalas-Medici, R. Uses and abuses of the mediation construct: The case of operant reinforcement of autonomic and neural responses. *Behaviorism,* 1972, *1,* 103–117.

Geller, A., Jarvik, M. E., & Robustelli, F. Incubation and the Kamin effect. *Journal of Experimental Psychology,* 1970, *85,* 61–65.

Gibbon, J. Timing and discrimination of shock density in avoidance. *Psychological Review,* 1972, *79,* 68–92.

Gilbert, R. M., & Sutherland, N. S. (Eds.), *Animal discrimination learning.* New York: Academic Press, 1969.

Glanzer, M. Curiosity, exploratory drive, and stimulus satiation. *Psychological Bulletin,* 1958, *55,* 302–315.

Goodrich, K. P. Running speed and drinking rate as functions of sucrose concentrations and amount of consummatory activity. *Journal of Comparative and Physiological Psychology,* 1960, *53,* 245–250.

Gormezano, I. Investigations of defense and reward conditioning in the rabbit. In A. H. Black and W. F. Prokasy (Eds.), *Classical conditioning II: Current research and theory.* New York: Appleton-Century-Crofts, 1972.

Gormezano, I., & Moore, J. W. Effects of instructional set and UCS intensity on the latency, percentage, and form of the eyelid response. *Journal of*

Experimental Psychology, 1962, *63,* 487–494.

Grant, D. A., & Schipper, L. M. The acquisition and extinction of conditioned eyelid responses as a function of the percentage of fixed-ratio random reinforcement. *Journal of Experimental Psychology,* 1952, *43,* 313–320.

Grant, D. A., & Schneider, D. B. Intensity of the CS and strength of conditioning I. The conditioned eyelid response to light. *Journal of Experimental Psychology,* 1948, *38,* 690–696.

Greenspoon, J. The reinforcing effect of two spoken sounds on the frequency of two responses. *American Journal of Psychology,* 1955, *68,* 409–416.

Grice, G. R. The relation of secondary reinforcement to delayed reward in visual discrimination learning. *Journal of Experimental Psychology,* 1948, *38,* 1–16.

Grice, G. R. Investigations of response mediated generalization. In D. I. Mostofsky (Ed.), *Stimulus generalization.* Stanford, Calif.: Stanford University Press, 1965.

Grice, G. R., & Davis, J. D. Mediated stimulus equivalence and distinctiveness in human conditioning. *Journal of Experimental Psychology,* 1958, *55,* 565–571.

Grice, G. R., & Davis, J. D. Effect of concurrent responses on the evocation and generalization of the conditioned eyeblink. *Journal of Experimental Psychology,* 1960, *59,* 391–395.

Grossen, N. E., & Kelley, M. J. Species-specific behavior and acquisition of avoidance behavior in rats. *Journal of Comparative and Physiological Psychology,* 1972, *81,* 307–310.

Grossen, N. E., Kostansek, D. J., & Bolles, R. C. Effects of appetitive discriminative stimuli on avoidance behavior. *Journal of Experimental Psychology,* 1969, *81,* 340–343.

Guthrie, E. R. Reward and punishment. *Psychological Review,* 1934, *41,* 450–460.

Guttman, N., & Kalish, H. I. Discriminability and stimulus generalization. *Journal of Experimental Psychology,* 1956, *51,* 79–88.

Haas, R. B., Shessel, F. M., Willner, H. S., & Rescorla, R. A. The effect of satiation following partial reinforcement. *Psychonomic Science,* 1970, *18,* 296–297.

Hablitz, J. J., & Braud, W. G. Adrenalin, sodium amobarbital, and the Kamin effect in the albino rat. *Learning and Motivation,* 1972, *3,* 51–58.

Hall, J. F. Studies in secondary reinforcement: II. Secondary reinforcement as a function of drive during primary reinforcement. *Journal of Comparative and Physiological Psychology,* 1951, *44,* 462–466.

Hammond, L. J. Increased responding to CS in differential CER. *Psychonomic Science,* 1966, *5,* 337–338.

Hanson, H. M. Effects of discrimination training on stimulus generalization. *Journal of Experimental Psychology,* 1959, *58,* 321–334.

Hanson, H. M. Stimulus generalization following three-stimulus discrimination training. *Journal of Comparative and Physiological Psychology,* 1961, *54,* 181–185.

Harlow, H. F. The formation of learning sets. *Psychological Review,* 1949, *56,* 51–65.

Harlow, H. F. Analysis of discrimination learning by monkeys. *Journal of Experimental Psychology,* 1950, *40,* 26–39.

Harlow, H. F. Learning set and error factor theory. In S. Koch (Ed.), *Psychology: A study of a science,* Vol. 2. New York: McGraw-Hill, 1959.

Harlow, H. F., & Warren, J. M. Formation and transfer of discrimination learning sets. *Journal of Comparative and Physiological Psychology,* 1952, *45,* 482–489.

Hastings, S. E., & Obrist, P. A. Heart rate during conditioning in humans: Effect of varying the interstimulus (CS-UCS) interval. *Journal of Experimental Psychology,* 1967, *74,* 431–442.

Hearst, E., & Koresko, M. B. Stimulus generalization and amount of prior training on variable-interval reinforcement. *Journal of Comparative and Physiological Psychology,* 1968, *66,* 133–138.

Hebert, J. A., & Krantz, D. L. Transposition: A reevaluation. *Psychological Bulletin,* 1965, *63,* 244–257.

Hendry, D. P. *Conditioned reinforcement.* Homewood, Ill.: Dorsey Press, 1969.

Herrnstein, R. J. Method and theory in the study of avoidance. *Psychological Review,* 1969, *76,* 49–69.

Herrnstein, R. J., & Hineline, P. N. Negative reinforcement as shock-frequency reduction. *Journal of the Experimental Analysis of Behavior,* 1966, *9,* 421–430.

Hill, W. F., & Spear, N. E. Resistance to extinction as a joint function of reward magnitude and the spacing of extinction trials. *Journal of Experimental Psychology,* 1962, *64,* 636–639.

Hoffman, H. S., & Fleshler, M. Aversive control with the pigeon. *Journal of the Experimental Analysis of Behavior,* 1959, *2,* 213–218.

Hovland, C. I. The generalization of conditioned responses: I. The sensory generalization of conditioned responses with varying frequencies of tone. *Journal of General Psychology,* 1937, *17,* 125–148. (a)

Hovland, C. I. The generalization of conditioned responses: II. The sensory generalization of conditioned responses with varying intensities of tone. *Journal of Genetic Psychology,* 1937, *51,* 279–291. (b)

Hull, C. L. *Principles of behavior.* New York: Appleton-Century-Crofts, 1943.

Hull, C. L. Stimulus intensity dynamism (V) and stimulus generalization. *Psychological Review,* 1949, *56,* 67–76.

Hulse, S. H. Amount and percentage of reinforcement and duration of goal confinement in conditioning and extinction. *Journal of Experimental Psychology,* 1958, *56,* 48–57.

Humphreys, L. G. The effect of random alternation of reinforcement on the acquisition and extinction of conditioned eyelid reactions. *Journal of Experimental Psychology,* 1939, *25,* 141–158.

Hunt, H. F., & Brady, J. V. Some effects of punishment and intercurrent "anxiety" on a simple operant. *Journal of Comparative and Physiological Psychology,* 1955, *48,* 305–310.

Ison, J. R. Experimental extinction as a function of number of reinforcements. *Journal of Experimental Psychology,* 1962, *64,* 314–317.

Ison, J. R., & Cook, P. E. Extinction performance as a function of incentive magnitude and number of acquisition trials. *Psychonomic Science*, 1964, *1*, 245–246.

Jenkins, H. M., & Harrison, R. H. Effect of discrimination training on auditory generalization. *Journal of Experimental Psychology*, 1960, *59*, 246–253.

Jenkins, W. O. A temporal gradient of derived reinforcement. *American Journal of Psychology*, 1950, *63*, 237–243.

Jensen, A. R. On the reformulation of inhibition in Hull's system. *Psychological Bulletin*, 1961, *58*, 274–298.

Jensen, G. D., & Cotton, J. W. Running speed as a function of stimulus similarity and number of trials. *Journal of Comparative and Physiological Psychology*, 1961, *54*, 474–476.

Jernstedt, G. C. Joint effects of pattern of reinforcement, intertrial interval, and amount of reinforcement in the rat. *Journal of Comparative and Physiological Psychology*, 1971, *75*, 421–429.

Johnston, J. M. Punishment of human behavior. *American Psychologist*, 1972, *27*, 1033–1054.

Kalat, J. W., & Rozin, P. Role of interference in taste-aversion learning. *Journal of Comparative and Physiological Psychology*, 1971, *77*, 53–58.

Kalish, H. I. The relationship between discriminability and stimulus generalization: A reevaluation. *Journal of Experimental Psychology*, 1958, *55*, 637–644.

Kalish, H. I. Stimulus generalization. In M. H. Marx (Ed.), *Learning: Processes*. New York: Macmillan, 1969.

Kalish, H. I., & Haber, A. Generalization: I. Generalization gradients from single and multiple stimulus points. II. Generalization of inhibition. *Journal of Experimental Psychology*, 1963, *65*, 176–181.

Kalish, H. I., & Haber, A. Prediction of discrimination from generalization following variations in deprivation level. *Journal of Comparative and Physiological Psychology*, 1965, *60*, 125–128.

Kamin, L. J. Traumatic avoidance learning: The effects of CS-US interval with a trace conditioning procedure. *Journal of Comparative and Physiological Psychology*, 1954, *47*, 65–72.

Kamin, L. J. The effects of termination of the CS and avoidance of the US on avoidance learning. *Journal of Comparative and Physiological Psychology*, 1956, *49*, 420–424.

Kamin, L. J. The gradient of delay of secondary reward in avoidance learning. *Journal of Comparative and Physiological Psychology*, 1957, *50*, 445–449. (a)

Kamin, L. J. The gradient of delay of secondary reward in avoidance learning tested on avoidance trials only. *Journal of Comparative and Physiological Psychology*, 1957, *50*, 450–456. (b)

Kamin, L. J. The retention of an incompletely learned avoidance response. *Journal of Comparative and Physiological Psychology*, 1957, *50*, 457–460. (c)

Kamin, L. J. The delay-of-punishment gradient. *Journal of Comparative and Physiological Psychology,* 1959, *52,* 434–437.

Kamin, L. J. Predictability, surprise, attention, and conditioning. In B. A. Campbell and R. M. Church (Eds.), *Punishment and aversive behavior.* New York: Appleton-Century-Crofts, 1969.

Kamiya, J. Operant control of the EEG alpha rhythm and some of its reported effects on consciousness. In C. Tart (Ed.), *Altered states of consciousness.* New York: Wiley, 1969.

Karsh, E. B. Effects of number of rewarded trials and intensity of punishment on running speed. *Journal of Comparative and Physiological Psychology,* 1962, *55,* 44–51.

Karsh, E. B. Punishment: Trial spacing and shock intensity as determinants of behavior in a discrete operant situation. *Journal of Comparative and Physiological Psychology,* 1964, *58,* 299–302.

Katkin, E. S., & Murray, E. N. Instrumental conditioning of autonomically mediated behavior: Theoretical and methodological issues. *Psychological Bulletin,* 1968, *70,* 52–68.

Kaufman, A., & Baron, A. Conditioned reinforcing and aversive aspects of the stimuli defining the components of a two-component chain. *Genetic Psychology Monographs,* 1969, *80,* 151–201.

Keehn, J. D. On the nonclassical nature of avoidance behavior. *American Journal of Psychology,* 1959, *72,* 243–247.

Kelleher, R. T., & Fry, W. T. Stimulus functions in chained fixed-interval schedules. *Journal of the Experimental Analysis of Behavior,* 1962, *5,* 167–173.

Kelleher, R. T., & Gollub, L. R. A review of positive conditioned reinforcement. *Journal of the Experimental Analysis of Behavior,* 1962, *5,* 543–597.

Keller, F. S. "Goodbye, teacher . . ." *Journal of Applied Behavior Analysis,* 1968, *1,* 79–89.

Keller, F. S., & Schoenfeld, W. N. *Principles of psychology.* New York: Appleton-Century-Crofts, 1950.

Kimble, G. A. Shock intensity and avoidance learning. *Journal of Comparative and Physiological Psychology,* 1955, *48,* 281–284.

Kimble, G. A. *Hilgard and Marquis' conditioning and learning.* (2nd ed.) New York: Appleton-Century-Crofts, 1961.

Kimmel, H. D. Instrumental inhibitory factors in classical conditioning. In W. F. Prokasy (Ed.), *Classical conditioning.* New York: Appleton-Century-Crofts, 1965.

Kimmel, H. D. Instrumental conditioning of autonomically mediated behavior. *Psychological Bulletin,* 1967, *67,* 337–345.

Kinnaman, K. J. Mental life of two *Macacus rhesus* monkeys in captivity. *American Journal of Psychology,* 1902, *13,* 98–148, 173–218.

Kinsman, R. A., & Bixenstine, V. E. Secondary reinforcement and shock termination. *Journal of Experimental Psychology,* 1968, *76,* 62–68.

Kintsch, W. Runway performance as a function of drive strength and magnitude of reinforcement. *Journal of Comparative and Physiological Psychology,* 1962, *55,* 882–887.

Kirby, R. H. Acquisition, extinction, and retention of an avoidance response in rats as a function of age. *Journal of Comparative and Physiological Psychology*, 1963, *56*, 158–162.

Kish, G. B. Learning when the onset of illumination is used as reinforcing stimulus. *Journal of Comparative and Physiological Psychology*, 1955, *48*, 261–264.

Klein, R. M. Intermittent primary reinforcement as a parameter of secondary reinforcement. *Journal of Experimental Psychology*, 1959, *58*, 423–427.

Klein, S. B. Adrenal-pituitary influence in reactivation of avoidance-learning memory in the rat after intermediate intervals. *Journal of Comparative and Physiological Psychology*, 1972, *79*, 341–359.

Klein, S. B., & Spear, N. E. Forgetting by the rat after intermediate intervals ("Kamin effect") as retrieval failure. *Journal of Comparative and Physiological Psychology*, 1970, *71*, 165–170.

Koteskey, R. L. A stimulus-sampling model of the partial reinforcement effect. *Psychological Review*, 1972, *79*, 161–171.

Kraeling, D. Analysis of amount of reward as a variable in learning. *Journal of Comparative and Physiological Psychology*, 1961, *54*, 560–565.

Kramer, T. J., & Rilling, M. Differential reinforcement of low rates: A selective critique. *Psychological Bulletin*, 1970, *74*, 225–254.

Krasner, L., & Ullmann, L. P. (Eds.) *Research in behavior modification*. New York: Holt, Rinehart and Winston, 1965.

Kremer, E. F. Truly random and traditional control procedures in CER conditioning in the rat. *Journal of Comparative and Physiological Psychology*, 1971, *76*, 441–448.

Kremer, E. F., & Kamin, L. J. The truly random control procedure: Associative or nonassociative effects in the rat. *Journal of Comparative and Physiological Psychology*, 1971, *74*, 203–210.

Kurtz, P. S., & Shafer, J. N. The interaction of UCS intensity and intertrial interval in avoidance learning. *Psychonomic Science*, 1967, *8*, 465–466.

Kushner, M. Desensitization of a posttraumatic phobia. In L. P. Ullmann and L. Krasner (Eds.), *Case studies in behavior modification*. New York: Holt, Rinehart and Winston, 1965.

Lacey, J. I., Smith, R. L., & Green, A. Use of conditioned autonomic responses in the study of anxiety. *Psychosomatic Medicine*, 1955, *17*, 208–217.

Lashley, K. S., & Wade, M. The Pavlovian theory of generalization. *Psychological Review*, 1946, *53*, 72–87.

Lawrence, D. H., & DeRivera, J. Evidence for relational discrimination. *Journal of Comparative and Physiological Psychology*, 1954, *47*, 465–471.

Lazarus, A. A. *Behavior therapy and beyond*. New York: McGraw-Hill, 1971.

Leach, D. A. Rat's extinction performance as a function of deprivation level during training and partial reinforcement. *Journal of Comparative and Physiological Psychology*, 1971, *75*, 317–323.

Lederhendler, I., & Baum, M. Mechanical facilitation of the action of response prevention (flooding) in rats. *Behavior Research and Therapy*, 1970, *8*, 43–48.

Leitenberg, H. Is time-out from positive reinforcement an aversive event? *Psychological Bulletin*, 1965, *64*, 428–441.

Leonard, D. W. Amount and sequence of reward in partial and continuous reinforcement. *Journal of Comparative and Physiological Psychology*, 1969, *67*, 204–211.

Leung, C. M., & Jensen, G. D. Shifts in percentage of reinforcement viewed as changes in incentive. *Journal of Experimental Psychology*, 1968, *76*, 291–296.

Levine, M. A. A model of hypothesis behavior in discrimination learning set. *Psychological Review*, 1959, *66*, 353–366.

Levine, M. A. Hypothesis behavior. In A. M. Schrier, H. F. Harlow, and F. Stollnitz (Eds.), *Behavior of nonhuman primates: Modern research trends*. New York: Academic Press, 1965.

Levine, M. A., Levinson, B., & Harlow, H. F. Trials per problem as a variable in the acquisition of discrimination learning set. *Journal of Comparative and Physiological Psychology*, 1959, *52*, 396–398.

Levine, S. UCS intensity and avoidance learning. *Journal of Experimental Psychology*, 1966, *71*, 163–164.

Levine, S., & England, S. J. Temporal factors in avoidance learning. *Journal of Comparative and Physiological Psychology*, 1960, *53*, 282–283.

Lewis, D. J. Acquisition, extinction, and spontaneous recovery as a function of percentage of reinforcement and intertrial intervals. *Journal of Experimental Psychology*, 1956, *51*, 45–53.

Lewis, D. J. Partial reinforcement: A selective review of the literature since 1950. *Psychological Bulletin*, 1960, *57*, 1–28.

Lichtenstein, P. E. Studies of anxiety: I. The production of a feeding inhibition in dogs. *Journal of Comparative and Physiological Psychology*, 1950, *43*, 16–29.

Liddell, H. S. The conditioned reflex. In F. A. Moss (Ed.), *Comparative psychology*. New York: Prentice-Hall, 1934.

Linton, H. B., & Miller, N. E. The effect of partial reinforcement on behavior during satiation. *Journal of Comparative and Physiological Psychology*, 1951, *44*, 142–148.

Logan, F. A. Decision making by rats: Delay versus amount of reward. *Journal of Comparative and Physiological Psychology*, 1965, *59*, 1–12.

LoLordo, V. M. Positive conditioned reinforcement from aversive situations. *Psychological Bulletin*, 1969, *72*, 193–203.

London, P. The end of ideology in behavior modification. *American Psychologist*, 1972, *27*, 913–920.

Low, L. A., & Low, H. I. Effects of CS-US interval length upon avoidance responding. *Journal of Comparative and Physiological Psychology*, 1962, *55*, 1059–1061. (a)

Low, L. A., & Low, H. I. Effects of variable versus fixed CS-US interval schedules upon avoidance responding. *Journal of Comparative and Physiological Psychology*, 1962, *55*, 1054–1058. (b)

Lubow, R. E. Latent inhibition: Effects of frequency of nonreinforced preexposure of the CS. *Journal of Comparative and Physiological Psychology*, 1965, *60*, 454–457.

Mackintosh, N. J., Little, L., & Lord, J. Some determinants of behavioral contrast in pigeons and rats. *Learning and Motivation*, 1972, *3*, 148–161.

Maier, N. R. F. *Frustration: The study of behavior without a goal.* New York: McGraw-Hill, 1949.

Maier, S. F., Seligman, M. E. P., & Solomon, R. L. Pavlovian fear conditioning and learned helplessness: Effects on escape and avoidance behavior of (a) the CS-US contingency and (b) the independence of the US and voluntary responding. In B. A. Campbell and R. M. Church (Eds.), *Punishment and aversive behavior.* New York: Appleton-Century-Crofts, 1969.

Malmo, R. B. Classical and instrumental conditioning with septal stimulation as reinforcement. *Journal of Comparative and Physiological Psychology*, 1965, *60*, 1–8.

Margolius, G. Stimulus generalization of an instrumental response as a function of the number of reinforced trials. *Journal of Experimental Psychology*, 1955, *49*, 105–111.

Martin, L. K., & Riess, D. Effects of US intensity during previous discrete delay conditioning on conditioned acceleration during avoidance extinction. *Journal of Comparative and Physiological Psychology*, 1969, *69*, 196–200.

Martin, R. C., & Melvin, K. B. Vicious circle behavior as a function of delay of punishment. *Psychonomic Science*, 1964, *1*, 415–416.

Marx, M. H. Resistance to extinction as a function of continuous or intermittent presentation of a training cue. *Journal of Experimental Psychology*, 1958, *56*, 251–255.

Marx, M. H., & Murphy, W. W. Resistance to extinction as a function of the presentation of a motivating cue in the start box. *Journal of Comparative and Physiological Psychology*, 1961, *54*, 207–210.

Mason, W. A., Blazek, N. C., & Harlow, H. F. Learning capacities of the infant rhesus monkey. *Journal of Comparative and Physiological Psychology*, 1956, *49*, 449–453.

Masserman, J. H. *Behavior and neurosis.* Chicago: University of Chicago Press, 1943.

McAllister, D. E., & McAllister, W. R. Incubation of fear: An examination of the concept. *Journal of Experimental Research in Personality*, 1967, *2*, 180–190.

McAllister, W. R., & McAllister, D. E. Increase over time in the stimulus generalization of acquired fear. *Journal of Experimental Psychology*, 1963, *65*, 576–582.

McAllister, W. R., McAllister, D. E., & Douglass, W. K. The inverse relationship between shock intensity and shuttlebox avoidance learning: A reinforcement explanation. *Journal of Comparative and Physiological Psychology*, 1971, *74*, 426–433.

McCain, G. Partial reinforcement effects following a small number of acquisition trials. *Psychonomic Monograph Supplement*, 1966, *1*, 251–270.

McCain, G. Reward magnitude and instrumental responses: Consistent and partial reward. *Psychonomic Science*, 1970, *19*, 139–141.

McHose, J. H., & Tauber, L. Changes in delay of reinforcement in simple instrumental conditioning. *Psychonomic Science*, 1972, *27*, 291–292.

McMichael, J. S., & Corey, J. R. Contingency management in an introductory psychology course produces better learning. *Journal of Applied Behavior Analysis*, 1969, *2*, 79–83.

Medin, D. L. Role of reinforcement in discrimination learning set in monkeys. *Psychological Bulletin*, 1972, *77*, 305–318.

Mednick, S. A., & Freedman, J. L. Stimulus generalization, *Psychological Bulletin*, 1960, *57*, 169–200.

Mellgren, R. L., & Ost, J. W. P. Transfer of Pavlovian differential conditioning to an operant discrimination. *Journal of Comparative and Physiological Psychology*, 1969, *67*, 390–394.

Melvin, K. B., & Martin, R. C. Facilitative effects of two modes of punishment on resistance to extinction. *Journal of Comparative and Physiological Psychology*, 1966, *62*, 491–494.

Mikulka, P. J., & Pavlik, W. B. Deprivation level, competing responses, and the PRE. *Psychological Reports*, 1966, *18*, 95–102.

Miles, R. C. The relative effectiveness of secondary reinforcers throughout deprivation and habit-strength parameters. *Journal of Comparative and Physiological Psychology*, 1956, *49*, 126–130.

Miles, R. C. Discrimination in the squirrel monkey as a function of deprivation and problem difficulty. *Journal of Experimental Psychology*, 1959, *57*, 15–19.

Miles, R. C. Discrimination-learning sets. In A. M. Schrier, H. F. Harlow, and F. Stollnitz (Eds.), *Behavior of nonhuman primates: Modern research trends*. New York: Academic Press, 1965.

Miller, N. E. Studies of fear as an acquirable drive: I. Fear as motivation and fear reduction as reinforcement in the learning of new responses. *Journal of Experimental Psychology*, 1948, *38*, 89–101.

Miller, N. E. Learning resistance to pain and fear: Effects of overlearning, exposure, and rewarded exposure in context. *Journal of Experimental Psychology*, 1960, *60*, 137–145.

Miller, N. E. Learning of visceral and glandular responses. *Science*, 1969, *163*, 434–445.

Miller, N. E., & Banuazizi, A. Instrumental learning by curarized rats of a specific visceral response, intestinal, or cardiac. *Journal of Comparative and Physiological Psychology*, 1968, *65*, 1–7.

Miller, N. E., & Carmona, A. Modification of a visceral response, salivation in thirsty dogs, by instrumental training with water reward. *Journal of Comparative and Physiological Psychology*, 1967, *63*, 1–6.

Miller, N. E., & Kessen, M. L. Reward effects of food via stomach fistula compared with those via mouth. *Journal of Comparative and Physiological Psychology*, 1952, *45*, 555–564.

Misanin, J. R., Campbell, B. A., & Smith, N. F. Duration of punishment and the delay of punishment gradient. *Canadian Journal of Psychology*, 1966, *20*, 407–412.

Mollenauer, S. O. Shifts in deprivation level: Different effects depending on amount of preshift training. *Learning and Motivation*, 1971, *2*, 58–66.

Moltz, H. Latent extinction and the fractional anticipatory response mechanism. *Psychological Review*, 1957, *64*, 229–241.

Montgomery, K. C. The role of exploratory drive in learning. *Journal of Comparative and Physiological Psychology,* 1954, *47,* 60–64.

Moore, J. W. Differential eyelid conditioning as a function of the frequency and intensity of auditory CSs. *Journal of Experimental Psychology,* 1964, *68,* 250–259.

Moore, J. W., & Gormezano, I. Effects of omitted versus delayed UCS on classical eyelid conditioning under partial reinforcement. *Journal of Experimental Psychology,* 1963, *65,* 248–257.

Mostofsky, D. *Stimulus generalization.* Stanford, Calif.: Stanford University Press, 1965.

Mowrer, O. H. On the dual nature of learning: A reinterpretation of "conditioning" and "problem-solving." *Harvard Educational Review,* 1947, *17,* 102–148.

Mowrer, O. H. *Learning theory and personality dynamics.* New York: Ronald Press, 1950.

Mowrer, O. H. *Learning theory and behavior.* New York: Wiley, 1960.

Mowrer, O. H., & Aiken, E. G. Contiguity vs. drive reduction in conditioned fear: Variations in conditioned and unconditioned stimuli. *American Journal of Psychology,* 1954, *67,* 26–38.

Mowrer, O. H., & Jones, H. M. Extinction and behavior variability as functions of effortfulness of task. *Journal of Experimental Psychology,* 1943, *33,* 369–386.

Mowrer, O. H., & Lamoreaux, R. R. Avoidance conditioning and signal duration—a study of secondary motivation and reward. *Psychological Monograph,* 1942, *54* (Whole No. 247).

Mowrer, O. H., & Solomon, L. N. Contiguity vs. drive-reduction in conditioned fear: The proximity and abruptness of drive-reduction. *American Journal of Psychology,* 1954, *67,* 15–26.

Moyer, K. E. Effect of delay between training and extinction on the extinction of an avoidance response. *Journal of Comparative and Physiological Psychology,* 1958, *51,* 116–118.

Moyer, K. E., & Korn, J. H. Effect of UCS intensity on the acquisition and extinction of an avoidance response. *Journal of Experimental Psychology,* 1964, *67,* 352–359.

Muenzinger, K. F. Motivation in learning: I. Electric shock for correct responses in the visual discrimination habit. *Journal of Comparative Psychology,* 1934, *17,* 267–277.

Muenzinger, K. F., Bernstone, A. H., & Richards, L. Motivation in learning: VIII. Equivalent amounts of electric shock for right and wrong responses in a visual discrimination habit. *Journal of Comparative Psychology,* 1938, *26,* 177–186.

Muenzinger, K. F., & Fletcher, F. M. Motivation in learning: VII. The effect of an enforced delay at the point of choice in the visual discrimination habit. *Journal of Comparative Psychology,* 1937, *23,* 383–392.

Muenzinger, K. F., & Newcomb, H. Motivation in learning: V. The relative effectiveness of jumping a gap and crossing an electric grid in a visual discrimination habit. *Journal of Comparative Psychology,* 1936, *21,* 95–104.

Muenzinger, K. F., & Wood, A. Motivation in learning: V. The function of punishment as determined by its temporal relation to the act of choice in the visual discrimination habit. *Journal of Comparative Psychology*, 1935, *20*, 95–106.

Murray, A. K., & Strandberg, J. M. Development of a conditioned positive reinforcer through removal of an aversive stimulus. *Journal of Comparative and Physiological Psychology*, 1965, *60*, 281–283.

Myers, J. L. Secondary reinforcement: A review of recent experimentation. *Psychological Bulletin*, 1958, *55*, 284–301.

Nash, A. N., Muczyk, J. P., & Vettori, F. L. The relative practical effectiveness of programmed instruction. *Personnel Psychology*, 1971, *24*, 397–418.

Norris, E. B., & Grant, D. A. Eyelid conditioning as affected by verbally induced inhibitory set and counter reinforcement. *American Journal of Psychology*, 1948, *61*, 37–49.

Osgood, C. E. *Method and theory in experimental psychology.* New York: Oxford University Press, 1953.

Overmier, J. B. Instrumental and cardiac indices of Pavlovian fear conditioning as a function of US duration. *Journal of Comparative and Physiological Psychology*, 1966, *62*, 15–20.

Overmier, J. B., & Seligman, M. E. P. Effects of inescapable shock upon subsequent escape and avoidance responding. *Journal of Comparative and Physiological Psychology*, 1967, *63*, 28–33.

Padilla, A. M. A few acquisition trials: Effects of magnitude and percent reward. *Psychonomic Science*, 1967, *9*, 241–242.

Padilla, A. M. Analysis of incentive and behavioral contrast in the rat. *Journal of Comparative and Physiological Psychology*, 1971, *75*, 464–470.

Pavlik, W. B., & Reynolds, W. F. Effects of deprivation schedule and reward magnitude on acquisition and extinction performance. *Journal of Comparative and Physiological Psychology*, 1963, *56*, 452–455.

Pavlov, I. P. *Conditioned reflexes.* Translated by G. V. Anrep. London: Oxford University Press, 1927.

Perin, C. T. A quantitative investigation of the delay-of-reinforcement gradient. *Journal of Experimental Psychology*, 1943, *32*, 37–51.

Perkins, C. C., & Weyant, R. G. The interval between training and test trials as a determiner of the slope of generalization gradients. *Journal of Comparative and Physiological Psychology*, 1958, *51*, 596–600.

Peterson, N. Effect of monochromatic rearing on the control of responding by wavelength. *Science*, 1962, *136*, 774–775.

Premack, D. Toward empirical behavior laws: I. Positive reinforcement. *Psychological Review*, 1959, *66*, 219–233.

Premack, D. Reversibility of the reinforcement relation. *Science*, 1961, *136*, 255–257.

Premack, D. Predictions of the comparative reinforcement values of running and drinking. *Science*, 1963, *139*, 1062–1063. (a)

Premack, D. Rate differential reinforcement in monkey manipulation. *Journal of the Experimental Analysis of Behavior*, 1963, *6*, 81–89. (b)

Premack, D. Reinforcement theory. In D. Levine (Ed.), *Nebraska Symposium on motivation.* Lincoln, Nebr.: University of Nebraska Press, 1965.

Prokasy, W. F., & Chambliss, D. J. Temporal conditioning: Negative results. *Psychological Reports,* 1960, *7,* 539–542.

Prokasy, W. F., & Hall, J. F. Primary stimulus generalization. *Psychological Review,* 1963, *70,* 310–322.

Prokasy, W. F., Hall, J. F., & Fawcett, J. T. Adaptation sensitization, forward and backward conditioning, and pseudoconditioning of the GSR. *Psychological Reports,* 1962, *10,* 103–106.

Prokasy, W. F., & Whaley, F. L. The intertrial interval in classical conditioning. *Journal of Experimental Psychology,* 1961, *62,* 560–564.

Prokasy, W. F., & Whaley, F. L. Intertrial interval range shift in classical eyelid conditioning. *Psychological Reports,* 1963, *12,* 55–58.

Pubols, B. H. Constant versus variable delay of reinforcement. *Journal of Comparative and Physiological Psychology,* 1962, *55,* 52–56.

Quinsey, V. L. Conditioned suppression with no CS-US contingency in the rat. *Canadian Journal of Psychology,* 1971, *25,* 69–82.

Rachlin, H., & Herrnstein, R. J. Hedonism revisited: On the negative law of effect. In B. A. Campbell and R. M. Church (Eds.), *Punishment and aversive behavior.* New York: Appleton-Century-Crofts, 1969.

Rashotte, M. E., & Surridge, C. T. Partial reinforcement and partial delay of reinforcement effects with 72-hour intertrial intervals and interpolated continuous reinforcement. *Quarterly Journal of Experimental Psychology,* 1969, *21,* 156–161.

Ratner, S. C. Reinforcing and discriminative properties of the click in a Skinner box. *Psychological Reports,* 1956, *2,* 332.

Razran, G. H. S. A quantitative study of weaning by a conditioned salivary technique (semantic conditioning). *Science,* 1939, *90,* 89–90.

Razran, G. H. S. The observable unconscious and the inferable conscious in current Soviet psychophysiology: Interoceptive conditioning, semantic conditioning, and the orienting reflex. *Psychological Review,* 1961, *68,* 81–147.

Reese, H. W. Discrimination learning set in rhesus monkeys. *Psychological Bulletin,* 1964, *61,* 321–340.

Renner, K. E. Influence of deprivation and availability of goal box cues on the temporal gradient of reinforcement. *Journal of Comparative and Physiological Psychology,* 1963, *56,* 101–104.

Renner, K. E. Delay of reinforcement: An historical review. *Psychological Bulletin,* 1964, *61,* 341–361.

Renner, K. E. Delay of reinforcement and resistance to extinction: A supplementary report. *Psychological Reports,* 1965, *16,* 197–198.

Rescorla, R. A. Inhibition of delay in Pavlovian fear conditioning. *Journal of Comparative and Physiological Psychology,* 1967, *64,* 114–120. (a)

Rescorla, R. A. Pavlovian conditioning and its proper control procedures. *Psychological Review,* 1967, *74,* 71–80. (b)

Rescorla, R. A. Pavlovian conditioned inhibition. *Psychological Bulletin,* 1969, *72,* 77–94.

Rescorla, R. A., & LoLordo, V. Inhibition of avoidance behavior. *Journal of*

Comparative and Physiological Psychology, 1965, *59,* 406–412.

Rescorla, R. A., & Solomon, R. L. Two-process learning theory: Relationships between Pavlovian conditioning and instrumental learning. *Psychological Review,* 1967, *74,* 151–182.

Rescorla, R. A., & Wagner, A. R. A theory of Pavlovian conditioning: Variations in the effectiveness of reinforcement and nonreinforcement. In A. H. Black and W. Prokasy (Eds.), *Classical conditioning II: Current research and theory.* New York: Appleton-Century-Crofts, 1972.

Restle, F. Toward a quantitative description of learning set data. *Psychological Review,* 1958, *65,* 77–91.

Reynierse, J. H., Weisman, R. G., & Denny, M. R. Shock compartment confinement during the intertrial interval in avoidance learning. *Psychological Record,* 1963, *13,* 403–406.

Riess, D. Pavlovian phenomena in conditioned acceleration: Spontaneous recovery. *Psychonomic Science,* 1971, *23,* 351–353.

Riopelle, A. J., & Moon, W. H. Problem diversity and familiarity in multiple-discrimination learning by monkeys. *Animal Behavior,* 1968, *16,* 74–78.

Robbins, D. Partial reinforcement: A selective review of the alleyway literature since 1960. *Psychological Bulletin,* 1971, *76,* 415–431.

Roberts, W. A. Resistance to extinction following partial and consistent reinforcement with varying magnitudes of reward. *Journal of Comparative and Physiological Psychology,* 1969, *67,* 395–400.

Rosenbaum, G. Stimulus generalization as a function of level of experimentally induced anxiety. *Journal of Experimental Psychology,* 1953, *45,* 35–43.

Ross, S. M., & Ross, L. E. Comparison of trace and delay classical eyelid conditioning as a function of interstimulus interval. *Journal of Experimental Psychology,* 1971, *91,* 165–167.

Rozin, P., & Kalat, J. W. Specific hungers and poison avoidance as adaptive specializations of learning. *Psychological Review,* 1971, *78,* 459–486.

Saltzman, I. J. Maze learning in the absence of primary reinforcement: A study of secondary reinforcement. *Journal of Comparative and Physiological Psychology,* 1949, *42,* 161–181.

Schneiderman, N., & Gormezano, I. Conditioning of the nictitating membrane of the rabbit as a function of CS-US interval. *Journal of Comparative and Physiological Psychology,* 1964, *57,* 188–195.

Schoenfeld, W. N. An experimental approach to anxiety, escape, and avoidance behavior. In P. H. Hock and J. Zubin (Eds.), *Anxiety.* New York: Grune and Stratton, 1950.

Schoenfeld, W. N. *Theory of reinforcement schedules.* New York: Appleton-Century-Crofts, 1970.

Schoenfeld, W. N., Antonitis, J. J., & Bersh, P. J. A preliminary study of training conditions necessary for secondary reinforcement. *Journal of Experimental Psychology,* 1950, *40,* 40–45.

Schrier, A. M. Comparison of two methods of investigating the effect of amount of reward on performance. *Journal of Comparative and Physiological Psychology,* 1958, *51,* 725–731.

Scobie, S. R. Interaction of an aversive Pavlovian conditional stimulus with aversively and appetitively motivated operants in rats. *Journal of Com-*

parative and Physiological Psychology, 1972, *79*, 171–188.

Seidel, R. J. A review of sensory preconditioning. *Psychological Bulletin*, 1959, *56*, 58–73.

Seligman, M. E. P. On the generality of the laws of learning. *Psychological Review*, 1970, *77*, 406–418.

Seligman, M. E. P., & Hager, J. *Biological boundaries of learning.* New York: Appleton-Century-Crofts, 1972.

Seligman, M. E. P., & Maier, S. F. Failure to escape traumatic shock. *Journal of Experimental Psychology*, 1967, *74*, 1–9.

Seward, J. P., & Levy, N. Sign learning as a factor in extinction. *Journal of Comparative and Physiological Psychology*, 1949, *39*, 660–668.

Sheffield, F. D., & Roby, T. B. Reward value of a nonnutritive sweet taste. *Journal of Comparative and Physiological Psychology*, 1950, *43*, 471–481.

Sheffield, V. F. Extinction as a function of partial reinforcement and distribution of practice. *Journal of Experimental Psychology*, 1949, *39*, 511–526.

Sheffield, V. F. Resistance to extinction as a function of the distribution of extinction trials. *Journal of Experimental Psychology*, 1950, *40*, 305–313.

Sidman, M. Avoidance conditioning with brief shock and no exteroceptive warning signal. *Science*, 1953, *118*, 157–158. (a)

Sidman, M. Two temporal parameters of the maintenance of avoidance behavior by the white rat. *Journal of Comparative and Physiological Psychology*, 1953, *46*, 253–261. (b)

Sidman, M. Some properties of the warning stimulus in avoidance behavior. *Journal of Comparative and Physiological Psychology*, 1955, *48*, 444–450.

Siegel, P. S., Melvin, K. B., & Wagner, J. D. Vicious circle behavior in the rat: Measurement problems visited again. *Journal of Comparative and Physiological Psychology*, 1971, *76*, 311–315.

Siegel, P. S., & Milby, J. B. Secondary reinforcement in relation to shock termination: Second chapter. *Psychological Bulletin*, 1969, *72*, 146–156.

Singh, P. J., Sakellaris, P. C., & Brush, F. R. Retention of active and passive avoidance responses tested in extinction. *Learning and otivation*, 1971, *2*, 305–323.

Skinner, B. F. *The behavior of organisms.* New York: Appleton-Century-Crofts, 1938.

Skinner, B. F. 'Superstition' in the pigeon. *Journal of Experimental Psychology*, 1948, *38*, 168–172.

Skinner, B. F. Are theories of learning necessary? *Psychological Review*, 1950, *57*, 193–216.

Skinner, B. F. Teaching machines. *Science*, 1958, *128*, 969–977.

Smith, M. C. CS-US interval and US intensity in classical conditioning of the rabbit's nictitating membrane responses. *Journal of Comparative and Physiological Psychology*, 1968, *66*, 679–687.

Smith, N. F., Misanin, J. R., & Campbell, B. A. Effect of punishment on extinction of an avoidance response: Facilitation or inhibition? *Psychonomic Science*, 1966, *4*, 271–272.

Smith, R. C., & Bowles, S. M. A bibliography of experimental extinction. *Psychological Reports*, 1971, *29*, 895–930.

Solomon, R. L. Punishment. *American Psychologist*, 1964, *19*, 237–253.

Solomon, R. L., & Brush, E. S. Experimentally derived conceptions of anxiety and aversion. In M. R. Jones (Ed.), *Nebraska symposium on motivation.* Lincoln, Nebr.: University of Nebraska Press, 1956.

Solomon, R. L., Kamin, L. J., & Wynne, L. C. Traumatic avoidance learning: The outcomes of several extinction procedures with dogs. *Journal of Abnormal and Social Psychology,* 1953, *48,* 291–302.

Solomon, R. L., & Turner, L. H. Discriminative classical conditioning in dogs paralyzed by curare can later control discriminative avoidance responses in the normal state. *Psychological Review,* 1962, *69,* 202–219.

Solomon, R. L., & Wynne, L. C. Traumatic avoidance learning: The principles of anxiety conservation and partial irreversibility. *Psychological Review,* 1954, *61,* 353–385.

Spear, N. E., Klein, S. B., & Riley, E. P. The Kamin effect as "state-dependent learning": Memory-retrieval failure in the rat. *Journal of Comparative and Physiological Psychology,* 1971, *74,* 416–425.

Spear, N. E., & Spitzner, J. H. Simultaneous and successive contrast effects of reward magnitude in selective learning. *Psychological Monographs,* 1966, *80* (Whole No. 618).

Spence, K. W. The nature of discrimination learning in animals. *Psychological Review,* 1936, *43,* 427–449.

Spence, K. W. The differential response in animals to stimuli varying within a single dimension. *Psychological Review,* 1937, *44,* 430–444.

Spence, K. W. The role of secondary reinforcement in delayed-reward learning. *Psychological Review,* 1947, *54,* 1–8.

Spence, K. W. *Behavior theory and conditioning.* New Haven: Yale University Press, 1956.

Spence, K. W., & Norris, E. B. Eyelid conditioning as a function of the intertrial interval. *Journal of Experimental Psychology,* 1950, *40,* 716–720.

Spooner, A., & Kellogg, W. N. The backward conditioning curve. *American Journal of Psychology,* 1947, *60,* 321–334.

Staddon, J. E. R., & Simmelhag, V. L. The "superstition" experiment. A re-examination of its implications for the principles of adaptive behavior. *Psychological Review,* 1971, *78,* 3–43.

Stampfl, T. G., & Levis, D. J. Essentials of implosive therapy: A learning theory-based psychodynamic behavior therapy. *Journal of Abnormal Psychology,* 1967, *72,* 496–503.

Stanley, W. C. Extinction as a function of the spacing of extinction trials. *Journal of Experimental Psychology,* 1952, *43,* 249–260.

Staveley, H. E. Effect of escape duration and shock intensity on the acquisition and extinction of an escape response. *Journal of Experimental Psychology,* 1966, c72, 698–703.

Stein, L. Secondary reinforcement established with subcortical stimulation. *Science,* 1958, *127,* 466–467.

Storms, L. H., Boroczi, G., & Broen, W. E. Effects of punishment as a function of strain of rat and duration of shock. *Journal of Comparative and Physiological Psychology,* 1963, *56,* 1022–1026.

Strong, P. N. Memory for object discrimination in the rhesus monkey. *Journal of Comparative and Physiological Psychology,* 1959, *66,* 780–783.

Symmes, D., & Leaton, R. N. Failure to observe reinforcing properties of sound onset in rats. *Psychological Reports,* 1962, *10,* 458.

Tarpy, R. M. Reinforcement difference limen (RDL) for delay in shock escape. *Journal of Experimental Psychology,* 1969, *79,* 116–121.

Tarpy, R. M., & Koster, E. D. Stimulus facilitation of delayed-reward learning in the rat. *Journal of Comparative and Physiological Psychology,* 1970, *71,* 147–151.

Tarpy, R. M., & Sawabini, F. L. Reinforcement delay: A selective review of the last decade. *Psychological Bulletin,* 1974, *81,* 984–997.

Terrace, H. S. Discrimination training with and without "errors." *Journal of the Experimental Analysis of Behavior,* 1963, *6,* 1–27.

Terrace, H. S. Behavioral contrast and the peak shift: Effects of extended discrimination training. *Journal of the Experimental Analysis of Behavior,* 1966, *9,* 613–617. (a)

Terrace, H. S. Stimulus control. In W. K. Honig (Ed.), *Operant behavior: Areas of research and application.* New York: Appleton-Century-Crofts, 1966. (b)

Theios, J. The partial reinforcement effect sustained through blocks of continuous reinforcement. *Journal of Experimental Psychology,* 1962, *64,* 1–6.

Theios, J. Drive stimulus generalization increments. *Journal of Comparative and Physiological Psychology,* 1963, *56,* 691–695.

Theios, J., Lynch, A. D., & Lowe, W. F. Differential effects of shock intensity on one-way and shuttle avoidance conditioning. *Journal of Experimental Psychology,* 1966, *72,* 294–299.

Theios, J., & McGinnis, R. W. Partial reinforcement before and after continuous reinforcement. *Journal of Experimental Psychology,* 1967, *73,* 479–481.

Thistlethwaite, D. A critical review of latent learning and related experiments. *Psychological Bulletin,* 1951, *48,* 97–129.

Thomas, D. R. The effects of drive and discrimination training on stimulus generalization. *Journal of Experimental Psychology,* 1962, *64,* 24–28.

Thomas, D. R., & King, R. A. Stimulus generalization as a function of level of motivation. *Journal of Experimental Psychology,* 1959, *57,* 323–328.

Thomas, D. R., & Lopez, L. J. The effects of delayed testing on generalization slope. *Journal of Comparative and Physiological Psychology,* 1962, *55,* 541–544.

Thomas, D. R., & Mitchell, K. The role of instructions and stimulus categorizing in a measure of stimulus generalization. *Journal of the Experimental Analysis of Behavior,* 1962, *5,* 375–381.

Thomas, J. R. Fixed-ratio punishment by time-out of concurrent variable-interval behavior. *Journal of the Experimental Analysis of Behavior,* 1968, *11,* 609–616.

Thompson, R. F. Sensory preconditioning. In R. F. Thompson and J. S. Voss (Eds.), *Topics in learning and performance.* New York: Academic Press, 1972.

Thorndike, E. L. Animal intelligence. An experimental study of the associative processes in animals. *Psychological Monographs,* 1898, *2,* No. 8.

Thorndike, E. L. *The fundamentals of learning.* New York: Teachers College, 1932.

Tolman, E. C. *Purposive behavior in animals and men.* New York: Appleton-Century-Crofts, 1932.

Tolman, E. C., & Honzik, C. H. Introduction and removal of reward, and maze performance in rats. *University of California Publications of Psychology,* 1930, *4,* 257–275.

Tombaugh, T. N. Resistance to extinction as a function of the interaction between training and extinction delays. *Psychological Reports,* 1966, *19,* 791–798.

Tombaugh, T. N. Secondary reinforcement and the partial reinforcement effect in the rat. *Journal of Comparative and Physiological Psychology,* 1970, *71,* 160–164.

Tombaugh, T. N., & Tombaugh, J. W. Effects of delay of reinforcement and cues upon acquisition and extinction performance. *Psychological Reports,* 1969, *25,* 931–934.

Tombaugh, T. N., & Tombaugh, J. W. Effects on performance of placing a visual cue at different temporal locations within a constant delay interval. *Journal of Experimental Psychology,* 1971, *87,* 220–224.

Tracy, W. K. Wavelength generalization and preference in monochromatically reared ducklings. *Journal of the Experimental Analysis of Behavior,* 1970, *13,* 163–178.

Trapold, M. A., & Fowler, H. Instrumental escape performance as a function of the intensity of noxious stimulation. *Journal of Experimental Psychology,* 1960, *60,* 323–326.

Trapold, M. A., & Spence, K. W. Performance changes in eyelid conditioning as related to the motivational and reinforcing properties of the UCS. *Journal of Experimental Psychology,* 1960, *59,* 209–213.

Trapold, M. A., & Winokur, S. Transfer from classical conditioning and extinction to acquisition, extinction, and stimulus generalization of a positively reinforced instrumental response. *Journal of Experimental Psychology,* 1967, *73,* 517–525.

Turner, L. H., & Solomon, R. L. Human traumatic avoidance learning: Theory and experiments on the operant-respondent distinction and failures to learn. *Psychological Monographs,* 1962, *76* (Whole No. 559).

Tyler, D. W., Wortz, E. C., & Bitterman, M. E. The effect of random and alternating partial reinforcement on resistance to extinction in the rat. *American Journal of Psychology,* 1953, *66,* 57–65.

Van-Toller, C., & Tarpy, R. M. Immunosympathectomy and avoidance behavior. *Psychological Bulletin,* 1974, *81,* 132–137.

Verplanck, W. S. Unaware of where's awareness: Some verbal operants—Notates, Monents, and Notants. In C. W. Erikson (Ed.), *Behavior and awareness.* Durham: Duke University Press, 1962.

Wagner, A. R. Effects of amount and percentage of reinforcement and number of acquisition trials on conditioning and extinction. *Journal of Experimental Psychology,* 1961, *62,* 234–242.

Wagner, A. R. Stimulus selection and a "modified continuity theory." In G. H. Bower and J. T. Spence (Eds.), *The psychology of learning and motivation,* Vol. 3. New York: Academic Press, 1969.

Walker, E. L. Reinforcement—"the one ring." In J. T. Tapp (Ed.), *Reinforcement and behavior.* New York: Academic Press, 1969.

Warren, J. M. Solution of object and positional discriminations by rhesus monkeys. *Journal of Comparative and Physiological Psychology,* 1959, *52,* 92–93.

Warren, J. M. Primate learning in comparative perspective. In A. M. Schrier, H. F. Harlow, and F. Stollnitz (Eds.), *Behavior of nonhuman primates: Modern research trends.* New York: Academic Press, 1965.

Weinstock, R. B. Maintenance schedules and hunger drive: An examination of the rat literature. *Psychological Bulletin,* 1972, *78,* 311–320.

Weinstock, S. Resistance to extinction of a running response following partial reinforcement under widely spaced trials. *Journal of Comparative and Physiological Psychology,* 1954, *47,* 318–322.

Weiss, J. M. Effects of coping responses on stress. *Journal of Comparative and Physiological Psychology,* 1968, *65,* 251–260.

Weiss, J. M. Effects of coping behavior in different warning signal conditions on stress pathology in rats. *Journal of Comparative and Physiological Psychology,* 1971, *77,* 1–13.

Wenger, N., & Zeaman, D. Strength of cardiac CR's with varying unconditioned stimulus durations. *Psychological Review,* 1958, *65,* 238–241.

Wenrick, W. W. *A primer of behavior modification.* Belmont, Calif.: Brooks/ Cole, 1970.

White, C. T., & Schlosberg, H. Degree of conditioning of the GSR as a function of the period of delay. *Journal of Experimental Psychology,* 1952, *43,* 357–362.

Wickens, D. D. A study of voluntary and involuntary finger conditioning. *Journal of Experimental Psychology,* 1939, *25,* 127–140.

Wickens, D. D., Schroder, H. M., & Snide, J. D. Primary stimulus generalization of the GSR under two conditions. *Journal of Experimental Psychology,* 1954, *47,* 52–56.

Wike, E. L. *Secondary reinforcement: Selected experiments.* New York: Harper & Row, 1966.

Wike, E. L. Secondary reinforcement: Some research and theoretical issues. In W. J. Arnold and D. Levine (Eds.), *Nebraska symposium on motivation.* Lincoln, Nebr.: University of Nebraska Press, 1969.

Wike, E. L., & Farrow, B. J. The effects of drive intensity on secondary reinforcement. *Journal of Comparative and Physiological Psychology,* 1962, *55,* 1020–1023.

Wilcoxon, H. D. Historical introduction to the problem of reinforcement. In J. T. Tapp (Ed.), *Reinforcement and behavior.* New York: Academic Press, 1969.

Wilcoxon, H. C., Dragoin, W. B., & Kral, P. A. Illness-induced aversions in rat and quail: Relative salience of visual and gustatory cues. *Science,* 1971, *171,* 826–828.

Wilkes, W. P., & Crowder, W. F. Secondary reinforcement with and without control for response facilitation. *Journal of Psychology,* 1960, *49,* 83–86.

Williams, D. R., & Williams, H. Auto-maintenance in the pigeon: Sustained pecking despite contingent nonreinforcement. *Journal of the Experimental*

Analysis of Behavior, 1969, *12,* 511–520.

Williams, S. B. Resistance to extinction as a function of the number of reinforcements. *Journal of Experimental Psychology,* 1938, *23,* 506–522.

Wilson, J. J. Level of training and goal-box movements as parameters of the partial reinforcement effect. *Journal of Comparative and Physiological Psychology,* 1964, *57,* 211–213.

Wilson, R. S. Cardiac response: Determinants of conditioning. *Journal of Comparative and Physiological Psychology Monograph,* 1969, *68* (Pt. 2).

Wilson, W., Weiss, E. J., & Amsel, A. Two tests of the Sheffield hypothesis concerning resistance to extinction, partial reinforcement, and distribution of practice. *Journal of Experimental Psychology,* 1955, *50,* 51–60.

Wischner, G. J. The effect of punishment on discrimination learning in a noncorrection situation. *Journal of Experimental Psychology,* 1947, *37,* 271–284.

Wischner, G. J., & Fowler, H. Discrimination performance as affected by duration of shock for either the correct or incorrect response. *Psychonomic Science,* 1964, *1,* 239–240.

Wischner, G. J., Fowler, H., & Kushnick, S. A. Effect of strength of punishment for "correct" or "incorrect" responses on visual discrimination performance. *Journal of Experimental Psychology,* 1963, *65,* 131–138.

Wolfe, J. B. The effect of delayed reward upon learning in the white rat. *Journal of Comparative Psychology,* 1934, *17,* 1–21.

Wolpe, J. Experimental neuroses as learned behavior. *British Journal of Psychology,* 1952, *43,* 243–268.

Wolpe, J. Learning theory and "abnormal fixations." *Psychological Review,* 1953, *60,* 111–116.

Wolpe, J. *Psychotherapy by reciprocal inhibition.* Stanford: Stanford University Press, 1958.

Wyckoff, L. B., Sidowski, J., & Chambliss, D. J. An experimental study of the relationship between secondary reinforcing and cue effects of a stimulus. *Journal of Comparative and Physiological Psychology,* 1958, *51,* 103–109.

Wynne, L. C., & Solomon, R. L. Traumatic avoidance learning: Acquisition and extinction in dogs deprived of normal peripheral autonomic functioning. *Genetic Psychology Monograph,* 1955, *52,* 241–284.

Young, F. A. An attempt to obtain pupillary conditioning with infrared photography. *Journal of Experimental Psychology,* 1954, *48,* 62–68.

Zaretsky, H. H. Runway performance during extinction as a function of drive and incentive. *Journal of Comparative and Physiological Psychology,* 1965, *60,* 463–464.

Zaretsky, H. H. Learning and performance in the runway as a function of the shift in drive and incentive. *Journal of Comparative and Physiological Psychology,* 1966, *62,* 218–221.

Zimmerman, D. W. Durable secondary reinforcement. Method and theory. *Psychological Review,* 1957, *64,* 373–383.

Zimmerman, D. W. Sustained performance in rats based on secondary reinforcement. *Journal of Comparative and Physiological Psychology,* 1959, *52,* 353–358.

Index

Author Index

Subject Index

US = Unconditioned Stimulus
UR = Unconditioned Response
CS = Conditioned Stimulus
CR = Conditioned Response

Acquisition: definition of, 11; test of secondary reinforcement, 167–169

Aftereffects, and partial reinforcement effect, 157–160

Age, and retention, 109

Associationists, 7–8

Attention: in classical conditioning, 39; and punished discrimination, 124–125

Autonomic responses: in avoidance, 100–101; in classical conditioning, 33–34; in instrumental conditioning, 68–71

Autoshaping, 214–216

Avoidance training: and autonomic conditioning, 69; and CS offset, 93–96; and CS-US interval, 106–108; definition of, 48; and extinction, 144–148; and fear, 91–96; and intertrial interval, 108–109; and Pavlovian CS, 82–84; and preparedness, 217; prior shock and, 111–112; and punishment, 127–132; retention of, 109–111; and shock frequency reduction, 100; Sidman, 96–97; and species specific reactions, 102–104, 214–215; after sympathectomy, 101; theories of, 93–104; and US intensity, 104–106; and vicious circle, 131–132

Backward conditioning, 20, 22, 31

Behavior modification, 220–222

Brain stimulation: as reward, 69; as US, 78

Chaining, and secondary reinforcement, 164–169

Classical conditioning: appetitive, 15–16; of autonomic responses, 33; with compound CS, 37–39; defense, 16; and instrumental conditioning, 50, 73, 90–91; of internal responses, 33–34; models of, 11–12; operations of, 18–19; procedures for, 19–20; terms of, 16–18; theories of, 36–39

Competing responses: and avoidance retention, 110–111; and flooding, 147–148; in punishment, 119–120

Conditioned emotional response,- 39, 80–81; and punishment, 117, 119; and shock intensity, 81

Conditioned response: definition of, 17–18; effect and extinction, 141; in instrumental conditioning, 46; similarity to UR, 76–77; specification of, 20–23

Conditioned stimulus: compound, 37–39; definition of, 17; in instrumental conditioning, 45–46; intensity of, 29; offset in avoidance, 94–96; /US correlation, 37; /US interval in avoidance, 106–108; /US interval in classical conditioning, 26–27; /US interval in secondary reinforcement, 174; /US interval in taste aversion, 34–36

Consciousness, 5–6, 8

Consummatory behavior, and punishment, 137

Contiguity: in classical conditioning, 36–39; hypothesis of secondary reinforcement, 174, 175–178; Law of, 8, 16; in taste aversion, 34–36

Contingency: in instrumental conditioning, 43–45; in punishment, 116–118

Contrast: and amount of reinforcement, 50–54; and incentive, 53–54; and reward delay, 56; and re-

ward frequency, 52
Curare: in autonomic conditioning, 69–71; in classical conditioning, 37; and helplessness, 112; in sensory preconditioning, 32

Darwin, 8–9
Delay: of CS offset, 93–96; of punishment, 122, 129–131; of reward, 54–56, 60–62; of reward and extinction, 142–144; of reward and secondary reinforcement, 171–172
Delayed conditioning: and avoidance, 99–100, 106; definition of, 20
Discrimination: definition of, 198–199; difficulty and punishment, 126–127; and generalization, 189–190, 115–197; learning sets, 200–204; methods, 199; and partial reinforcement effect, 154–156; and peak shift, 195–197; and prior classical conditioning, 85; and punishment, 124–127; theories of, 199; and vicious circle, 133–135
Discriminative stimulus: in avoidance, 93, 96–98; definition of, 45–46; hypothesis of secondary reinforcement, 174–178
Disinhibition, 25–26
Distinctions between classical and instrumental conditioning, 73–75; operational, 75; reinforcement, 77–79; response, 75–77
Drive: acquired, 91–92; and extinction, 144; and partial reinforcement effect, 153; reduction theory, 208–210; and secondary reinforcement, 170

Emotion, 18, 79–80, 91–92
Escape training: definition of, 48; and reward amount, 59–60; and reward delay, 60–62; and US intensity, 58–59; and vicious circle, 131
Evolution: and avoidance, 102–103; influence of, 8–9; and preparedness, 216–218

Extinction: definition of, 11, 43, 140–141; and drive level, 58, 144; and fear, 144–148; and inhibition, 23–24; and intertrial interval, 144; after partial reinforcement, 151–160; and prior acquisition training, 142; and response effect, 141; and reward amount, 141–142; and reward delay, 142–144; test of secondary reinforcement, 167; theories of, 148–151

Fear: and avoidance, 93–96; and avoidance retention, 110–111; in classical conditioning, 18, 79–80, 91–92; conservation of, 145; and extinction, 144–148; and neurosis, 135–137; and punishment, 120–121; and vicious circle, 132–133
Fixation, and punishment, 135–137
Flooding, 145–148
Frustration: and partial reinforcement effect, 156; and punishment, 136; theory in extinction, 148–151

Generalization: definition of, 184–185; and early experience, 194–195; mediated, 190–192; and motivation, 197–198; primary, 185–187; and prior discrimination, 195–197; semantic, 187–188; and test interval, 193–194; tests of, 185; theories of, 188–192; and training trials, 192–193

Helplessness, 112

Incentive: and contrast, 53–54; hypothesis of secondary reinforcement, 179–181; and r_g-s_g, 53–54
Information: and avoidance, 97–98; and classical conditioning, 37; hypothesis of secondary reinforcement, 178–179
Inhibition: and classical/instrumental interaction, 83–86; conditioned, 24; definition of, 23; of delay, 25–26, 85–86; in discrimination, 199–

200; and extinction, 23–24; and spontaneous recovery, 23–24; theory in extinction, 148–149

Instructions, 29–30

Instrumental conditioning: appetitive, 42–43; of autonomic responses, 68–71; aversive, 43; basic paradigms, 46–49; and classical conditioning, 50, 73, 90–91; operations, 43–45; terms, 45–46

Interaction between classical and instrumental conditioning, 79–87

Interoceptive conditioning, 33–34, 78

Intertrial interval: in avoidance, 108–109; in classical conditioning, 27; in escape, 60; and extinction, 144; and sequential theory, 157

Introspection, 5–6

Law of effect: and autonomic conditioning, 71; definition of, 43–44; and punishment, 116–117, 121; as theory, 210–213

Learning: definition of, 3–4; latent, 10–11; measurement of, 9–12; programmed, 218–220

Learning sets: definition of, 200–202; factors governing, 202–203; theories of, 203–204

Modified conditioning theory, 37–39

Motivation: and avoidance, 93–96; fear, 91–92; and generalization, 197–198; and incentive, 53–54

Neurosis: and behavior modification, 220–222; and punishment, 115–116

Omission training, 48

Orientation in sensory preconditioning, 31–32

Partial reinforcement effect: definition of, 151–152; and drive, 153; and patterning, 154; and reward magnitude, 152–153; and reward percentage, 152; and reward schedule shifts, 153–154; theories of, 154–160; and training level, 153

Peak shift, 195–197

Performance: indices of, 11; versus learning, 9–11

Preparedness, 216–218

Prepotent theory, 211–213

Pseudoconditioning, 21–22

Punishment: definition of, 48, 115; delay of, 122; and discrimination, 124–127; duration of, 122–123; and neurosis, 115–116, 135–137; prior, 123–124; theories of, 118–121; and US intensity, 121–122; and vicious circle, 131–135

Random control procedure, 22–23

Retention, of avoidance, 109–111

Reward training: amount of reinforcement in, 50; of autonomic responses, 69–70; definition of, 47; and drive level, 56–58; and incentive, 53–54; and Pavlovian CS, 84–85; and punishment, 121–127; and reward delay, 54–56; and shifts in reward amount, 50–54; and variable reward delay, 55

Schedules of reinforcement: in avoidance, 100; chained, 66; compound, 66; concurrent, 66; definition of, 62; and differential reinforcement of low rates of responding (DRL), 65–66; and extinction, 152–154; fixed interval, 64–65; fixed ratio, 62–63; mixed, 66; multiple, 66; in punishment, 117–118; and secondary reinforcement, 170–171; tandem, 66; variable interval, 65; variable ratio, 64

Secondary reinforcement: definition of, 163; and degree of training, 170; and drive, 170; experimental tests of, 164–169; and reward delay, 54–55; and reward magnitude, 171; and schedule of presentation, 172–173; and schedule of reward, 170–

2 3 4 5 6 7 8 9 10 –DB– 80 79 78 77 76 75